Management Technologies for Metal Mining Influenced Water

Sampling and Monitoring for the Mine Life Cycle

Volume 6

Edited by Virginia T. McLemore, Kathleen S. Smith, and Carol C. Russell

PUBLISHED BY THE
SOCIETY FOR MINING, METALLURGY & EXPLORATION

Society for Mining, Metallurgy & Exploration Inc. (SME)
12999 East Adam Aircraft Circle
Englewood, Colorado 80112
(303) 948-4200 / (800) 763-3132
www.smenet.org

SME advances the worldwide mining and minerals community through information exchange and professional development. SME is the world's largest association of mining and minerals professionals.

ISBN: 978-0-87335-355-7
Ebook: 978-0-87335-398-4

Library of Congress Cataloging-in-Publication Data
Sampling and monitoring for the mine life cycle / edited by Virginia T. McLemore, Kathleen S. Smith, and Carol C. Russell.
pages cm. -- (Management technologies for metal mining influenced water ; volume 6)
Includes bibliographical references and index.
ISBN 978-0-87335-355-7 (print) -- ISBN 978-0-87335-398-4 (ebook)
1. Mines and mineral resources--Environmental aspects. 2. Runoff. 3. Groundwater--Pollution. 4. Water--Sampling. 5. Environmental monitoring. I. McLemore, Virginia T., editor of compilation. II. Smith, Kathleen S. (Geologist), editor of compilation. III. Russell, Carol C., editor of compilation.
TD195.M5S258 2014
628.1'6832--dc23
2014005639

On the Cover: Richard B. Wanty (USGS) collecting water quality samples near the bottom of the Roaring River in Rocky Mountain National Park, Colorado. Photograph courtesy of Philip L. Verplanck (USGS).

DISCLAIMER

Contents

Preface xiii

Chapter 1	**Introduction**	**1**
	1.1 Synopsis	1
	1.2 Organization of This Handbook	1
	1.3 Role of this Handbook	2
	1.4 Importance of Defining Questions	3
	1.5 Sampling and Monitoring for Closure	4
	1.6 Terminology	4
Chapter 2	**Sampling and Monitoring During the Mining Phases**	**9**
	2.1 Introduction	9
	2.2 Exploration (Premining/Undisturbed)	9
	2.3 Mine Development	18
	2.4 Operations	21
	2.5 Closure/Postclosure	25
	2.6 Historical, Inactive, and Abandoned Mines	26
	2.7 Summary	28
Chapter 3	**Sampling Considerations in the Mining Environment**	**33**
	3.1 Introduction	33
	3.2 Sampling Overview	37
	3.3 Water Sampling Considerations	39
	3.4 Solid Sampling Considerations	52
	3.5 Drillhole/Borehole Sampling Considerations	70
	3.6 Biological Sampling Considerations—Aquatic	73
	3.7 Air Sampling Considerations	76
	3.8 Quality Assurance/Quality Control Considerations	77
	3.9 Analytical Chemistry Considerations	84
	3.10 Contamination Considerations	86
	3.11 Sample Preservation, Handling, and Storage Considerations	88
	3.12 Summary	88
	3.13 Internet Resources	90

Chapter 4 **Decision Making, Risk, and Uncertainty** **111**
4.1 Introduction **111**
4.2 Risk **112**
4.3 International Guidance Document Commonalities **122**
4.4 Summary **129**

Chapter 5 **The Planning Process** **133**
5.1 Introduction **133**
5.2 Define Questions and Objectives **133**
5.3 Develop Site Conceptual Models **136**
5.4 Conceptual Models for Numerical Modeling **139**
5.5 Summary **143**

Chapter 6 **Sampling and Monitoring Program Implementation** **145**
6.1 Introduction **145**
6.2 Sampling and Monitoring Plan **146**
6.3 Sampling Design and the Sampling Plan Document **149**
6.4 Conduct the Pilot Study **150**
6.5 Implement the Sampling and Monitoring Plan (Program Implementation) **151**
6.6 Summary **151**

Chapter 7 **Data Management, Assessment, and Analysis for Decision Making** **153**
7.1 Introduction **153**
7.2 Data Management **153**
7.3 Analysis and Assessment of Data **155**
7.4 Data Presentation **158**
7.5 Data Interpretation **160**
7.6 Decision-Making Process **164**
7.7 Summary **169**
7.8 Internet Resources **169**

Chapter 8 **Additional Key Issues and Future Research Needs** **171**
8.1 Introduction **171**
8.2 Slope Stability and Failure **171**
8.3 Tailings Impoundments **172**
8.4 Subsidence **173**
8.5 Mine Openings **174**
8.6 Climate Change **174**
8.7 Sampling and Monitoring in Other Countries **177**
8.8 Future Research Needs and Topics **177**

The following appendices are contained on the CD-ROM.

Appendix 1 Selected Online Resources for Sampling, Monitoring, and Analytical Chemistry Methods

Appendix 2 Summary of Selected ASTM Methods

Appendix 3 Summary of Field Sampling and Analytical Methods with Bibliography

Appendix 4 Examples of Sampling Plans and Quality Assurance Project Plans

Appendix 5 Case Studies of Sampling and Monitoring

Appendix 6 Applications and Examples of Geo-Environmental Models (GEMs) at Mine Sites

Index 181

Editorial Board and Contributors

Senior Editors and Primary Authors

Virginia T. McLemore, New Mexico Bureau of Geology and Mineral Resources (NMBGMR)
Kathleen S. Smith, U.S. Geological Survey (USGS)
Carol C. Russell, U.S. Environmental Protection Agency (USEPA)

ADTI-MMS Sampling and Monitoring Committee

David A. Bird, SRK Consulting
Virginia T. McLemore, NMBGMR
Thomas C. Moyer, Black & Veatch Special Projects Corp.
Douglas C. Peters, TUVERA Exploration, Inc.
Carol C. Russell, USEPA
Ron L. Schmiermund, Economic & Environmental Geochemistry, Inc.
Kathleen S. Smith, USGS
Dennis L. Turner, Arizona Department of Environmental Quality
Thomas R. Wildeman, Colorado School of Mines (CSM)
Richard T. Wilkin, USEPA

Contributing Authors

Subsections

Susan J. Caplan, Bureau of Land Management
Karl L. Ford, Bureau of Land Management
James J. Gusek, Sovereign Consulting Inc.
Lisa Bithell Kirk, Enviromin, Inc.
Andrew S. Todd, USGS

Sidebars

Christopher H. Gammons, Montana Tech of The University of Montana
Briant A. Kimball, USGS
Lisa Bithell Kirk, Enviromin, Inc.
Virginia T. McLemore, NMBGMR
Patsy B. Moran, Arcadis U.S., Inc.
David A. Nimick, USGS
Charles A. Ramsey, EnviroStat, Inc.
Carol C. Russell, USEPA
Ron L. Schmiermund, Economic and Environmental Geochemistry, Inc.
Kathleen S. Smith, USGS
Stephen J. Sutley, USGS
Richard B. Wanty, USGS
Thomas R. Wildeman, CSM

Appendix 1: Compiled by Kathleen S. Smith (USGS), Carol C. Russell (USEPA), and Virginia T. McLemore (NMBGMR)

Appendix 2: Compiled by Samuel Nunoo, Frederick Ennin, Kojo Anim, and Virginia T. McLemore (NMBGMR)

Appendix 3: Compiled by Virginia T. McLemore, Kathleen S. Smith, and Carol C. Russell

Section Authors: Charles N. Alpers (USGS), Laurie S. Balistrieri (USGS), Lynn A. Brandvold (NMBGMR), Charles H. Bucknam (Newmont Metallurgical Technologies), Barbara A. Butler (USEPA), Thomas P. Chapin (USGS), Michael Copeland (USEPA), James G. Crock (USGS), Angelique D. Diaz (USEPA), Nelia Dunbar (NMBGMR), Andrea L. Foster (USGS), B. Frey (NMBGMR), Richard K. Glanzman (Glanzman Geochemical LLC), Mary P. Goldade (USEPA), Lareina Guenzel (USEPA), Ibrahim Gundiler (NMBGMR), Karin Olson Hoal (JKTech Pty Ltd.), Briant A. Kimball (USGS), Karl Laumbach (Human Systems Research, Inc.), William Lipps (ITT/OI Analytical), Raina M. Maier (University of Arizona), Robert R. McDougal (USGS), Virginia T. McLemore (NMBGMR), Suzette A. Morman (USGS), Samuel Nunoo (Freeport-McMoRan), Douglas C. Peters (TUVERA Exploration, Inc.), Timberley M. Roane (University of Colorado, Denver), Carol C. Russell (USEPA), James A. Russell (Summit Data Services), Travis S. Schmidt (USGS), Ron L. Schmiermund (Economic & Environmental Geochemistry, Inc.), Bruce D. Smith (USGS), Kathleen S. Smith (USGS), Sandra Spence (USEPA), Andrew S. Todd (USGS), Richard T. Wilkin (USEPA), and Mike Wireman (USEPA)

Appendix 4: Compiled by Carol C. Russell and Virginia T. McLemore

Appendix 5: Compiled by Kathleen S. Smith, Virginia T. McLemore, and Carol C. Russell

Case Study Authors: Michael C. Amacher (U.S. Forest Service), John M. Besser (USGS), Lynn Brandvold (NMBGMR), Barbara A. Butler (USEPA), Stanley E. Church (USGS), James F. Coles (USGS), James R. Herring (USGS), Richard G. Kiah (USGS), Briant A. Kimball (USGS), Lisa Bithell Kirk (Enviromin, Inc.), Virginia T. McLemore (NMBGMR), Nadine M. Piatak (USGS), Travis S. Schmidt (USGS), Robert R. Seal II (USGS), Lisa L. Stillings (USGS), Katherine Walton-Day (USGS), and R. David Williams (Bureau of Land Management)

Appendix 6: Virginia T. McLemore (NMBGMR)

Reviewers

Main Text: Thomas C. Moyer (Black & Veatch Special Projects Corp.), Patsy B. Moran (Arcadis U.S., Inc.), and C. Doc Richardson (Tetra Tech)

Appendix 1: Mary P. Goldade (USEPA) and Thomas C. Moyer (Black & Veatch Special Projects Corp.)

Appendix 2: Jeffrey A. Farrar (Bureau of Reclamation) and William Lipps (ITT/OI Analytical)

Appendix 3: Mary P. Goldade (USEPA), Thomas R. Wildeman (CSM), and Bradley S. Van Gosen (USGS)

Appendix 4: Thomas C. Moyer (Black & Veatch Special Projects Corp.) and Mary P. Goldade (USEPA)

Appendix 5: Thomas C. Moyer (Black & Veatch Special Projects Corp.) and David Rathke (USEPA)

Appendix 6: Ron L. Schmiermund (Economic & Environmental Geochemistry, Inc.) and Robert R. Seal II (USGS)

Overall: ADTI-MMS Sampling and Monitoring Committee and ADTI-MMS Steering Committee and Membership

Special thanks to David Bird (SRK Consulting, Inc.), Charles Bucknam (Newmont Metallurgical Technologies), Douglas Peters (TUVERA Exploration, Inc.), Janet Slate (USGS), and Patti Tyler (USEPA).

Figures and photographs were provided by Virginia T. McLemore, Kathleen S. Smith, Carol C. Russell, Meghan Jackson, Leo Galbadon, Tom Kaus, and Sharon F. Diehl.

Preface

Sampling and Monitoring for the Mine Life Cycle is volume 6 in the Management Technologies for Metal Mining Influenced Water series prepared through the Acid Drainage Technology Initiative–Metal Mining Sector (ADTI-MMS). Published handbooks in the series include *Basics of Metal Mining Influenced Water*; *Mitigation of Metal Mining Influenced Water*; *Mine Pit Lakes: Characteristics, Predictive Modeling, and Sustainability*; *Techniques for Predicting Metal Mining Influenced Water*; and *Geochemical Modeling for Mine Site Characterization and Remediation.* Each handbook undergoes a consensus review process that involves approval by the appropriate ADTI-MMS technical committee and ADTI-MMS membership.

ADTI-MMS is a technically focused consensus group of volunteer representatives from state and federal government, academia, the mining industry, consulting firms, and other interested parties who are involved in the environmentally sound management of metal mine wastes and drainage quality issues. ADTI-MMS is part of the Acid Drainage Technology Initiative (ADTI), which is the United States' regional member of the Global Alliance sponsored by the International Network for Acid Prevention (INAP). The mission of ADTI-MMS is to identify, evaluate, develop, and disseminate information about cost-effective, environmentally sound methods and technologies to manage mine wastes and related metallurgical materials for abandoned, active, and future mining and associated operations, and to promote understanding of these technologies. Drainage quality issues are considered for ore and waste as well as from alkaline, neutral, and acidic conditions in the mining and related metallurgical processing environment. ADTI-MMS does not address policy or regulatory matters. The co-chairs of the ADTI-MMS Sampling and Monitoring Technical Committee during the preparation of this handbook were Virginia T. McLemore (New Mexico Bureau of Geology and Mineral Resources [NMBGMR]), Kathleen S. Smith (U.S. Geological Survey [USGS]), and Carol C. Russell (U.S. Environmental Protection Agency [USEPA]).

This technical handbook is intended for planners, regulators, consultants, land managers, researchers, students, stakeholders, and anyone with an interest in mining influenced waters. The aim of this handbook is to provide specialized information applicable to sampling and monitoring activities at mining sites. This handbook is intended to supplement and enhance existing sampling and monitoring publications, and is not intended to provide policy or regulatory information.

This handbook greatly benefited from contributions and reviews by numerous experts listed in the front material. We thank them all for their hard work and patience during the preparation of this handbook. We extend special thanks to Charles Bucknam, Linda Figueroa, Jane Olivier, Doug Peters, and Dennis Turner for their leadership, support, encouragement, and tenacity. The NMBGMR (New Mexico Institute of Mining and Technology), U.S. Office of Surface Mining Reclamation and Enforcement (initial funding for report writing and obtaining MEND documents), and USEPA (travel funds) provided direct funding in support of this handbook. We acknowledge the in-kind support provided by companies, institutions, and agencies for contributors and reviewers to participate in the preparation of this handbook. We especially thank the NMBGMR, USGS (Mineral Resources Program and Toxic Substances Hydrology Program), and USEPA (Region 8 and the Office of Research and Development) for providing salary support for this work.

CHAPTER 1

Introduction

Kathleen S. Smith and Virginia T. McLemore

1.1 SYNOPSIS

This handbook provides an overview of basic aspects of sampling for environmental purposes and monitoring of environmentally relevant variables at mining sites. While focusing mainly on environmental sampling and monitoring of surface water, also considered are groundwater, process water streams, rock (native, mined, and processed), soil, and other media including air and biological organisms. The objectives of this handbook are to

- Endorse the principle that sampling and monitoring programs be designed to
  - Account for all aspects of the mine life cycle, and
  - Plan for closure throughout the mine life cycle;
- Discuss considerations for environmental sampling and monitoring at mining sites during the various phases of mining;
- Supply technical information and suggest resources that relate to sampling and monitoring at mining sites; and
- Provide an approach for and guidance on design of sampling and monitoring programs at mining sites.

This handbook will be useful to mine managers and planners, laboratory staff, consultants, regulators, researchers, undergraduate and graduate students, communities affected by mining activities, nongovernmental organizations, and the general public.

This is a technical document and is not intended to be used for policy or regulatory purposes. Furthermore, this handbook is not designed to be an all-inclusive compendium of information about sampling and monitoring of acidic drainage and other mining influenced water (MIW; term coined by Schmiermund and Drozd 1997) or natural drainage. Instead, it provides key information, technical details, and resources that can assist readers in making informed decisions about sampling and monitoring at mining sites. This handbook is intended to supplement and enhance current literature, and strives to provide an awareness of the critical components and complexities involved in environmental sampling and monitoring at mining sites. It differs from most information sources by providing an approach to address all types of MIW and other sampling media throughout the mine life cycle.

1.2 ORGANIZATION OF THIS HANDBOOK

This handbook is organized into a main text and six appendices. Sidebars and illustrations are included throughout the main text to provide additional detail about important concepts, to present examples and brief case studies, and to suggest resources for further information. References cited are listed at the end of each chapter.

The main text includes topical chapters that discuss various aspects of environmental sampling and monitoring at mining sites.

- Chapter 2 describes sampling and monitoring needs during the mine life cycle. The phases of mining are described, and the issues, challenges, sampling, monitoring, and characterization needs are specified for each mining phase.
- Chapter 3 discusses sampling considerations that relate to mining sites. Specialized sampling issues and relevant research findings for mining sites are described, and additional resources are suggested for further information.
- Chapter 4 provides an overview of risk and of international sampling and monitoring guidance documents. A stepwise framework is presented for developing and implementing a sampling and monitoring program at mining sites.
- Chapter 5 describes the planning process for sampling and monitoring at mining sites. It includes a discussion of the importance of conceptual models in this process.
- Chapter 6 covers the various aspects of implementing a sampling program at mining sites.
- Chapter 7 provides an overview of the decision-making process. It includes a discussion of the steps involved in decision making and highlights some of the specialized considerations for mining sites.
- Chapter 8 discusses key issues and research topics that need to be pursued at mining sites.

Appendices are included to supplement the main text. The appendices are available on the included CD-ROM and are an integral part of this handbook.

- Appendix 1 is a list of Web-site resources that address sampling and monitoring topics.
- Appendix 2 is a summary specifically devoted to ASTM standard methods that pertain to sampling and monitoring at mining sites.
- Appendix 3 contains technical summaries written by subject-matter experts that describe field measurements, collection methods, and analytical techniques and procedures, and discuss principles, limitations, and uses of various sampling and monitoring techniques. This appendix is designed to be a valuable resource to supplement concepts and methods discussed in the main text.
- Appendix 4 contains some examples of historical sampling plans for mining-related sites and also includes a USEPA (U.S. Environmental Protection Agency) Region 8 review form for Quality Management Plans.
- Appendix 5 is a collection of case studies from mining-related sites that demonstrate sampling and monitoring applications and discuss lessons learned. The case studies are written by scientists, and illustrate concepts and methods discussed in this handbook.
- Appendix 6 illustrates applications of GEMs (geo-environmental models) at mining sites using case histories.

1.3 ROLE OF THIS HANDBOOK

At present, there are few authoritative guides available that address environmental sampling and monitoring approaches for MIW throughout the mine life cycle. The Mine Environment Neutral Drainage (MEND) program has worked to develop technologies to prevent and control acidic drainage since 1989 and has published numerous significant reports concerning MIW. The Acid

Drainage Technology Initiative–Coal Mining Sector has produced two handbooks on acid mine drainage (AMD) as it relates to coal mining (Skousen et al. 1998; Kleinmann 2000). Since the 1980s, numerous general sampling and monitoring techniques have been developed throughout the world by USEPA, USGS (U.S. Geological Survey), and other organizations (TAEM and SENES 1997; USBR 2001; Tremblay and Hogan 2001a, 2001b; Popeck 2003; Downing 2010). The *Global Acid Rock Drainage Guide* (GARD Guide; INAP 2009) includes a section on monitoring but does not describe sampling methods and sampling strategy in detail, except briefly in some of the individual chapters. Hence, there are limited descriptions of the specialized approach necessary for environmental sampling and monitoring at mining sites. This handbook endeavors to fill that niche by providing specialized information that addresses sampling of all types of MIW and other sampling media throughout the mine life cycle.

1.4 IMPORTANCE OF DEFINING QUESTIONS

A recurring theme throughout this handbook is the importance of defining the questions being addressed and the objectives of the investigation. Prior to beginning a sampling and monitoring program, it is essential to know what data and information are necessary, how they are to be used, what sampling media will target which issues, the important analytes, and the desired degree of confidence in the answer(s). Sampling and monitoring decisions cannot be made, and protocols cannot be prepared, until these issues have been addressed. Various aspects of question definition are discussed later in this handbook.

It is good practice to think beyond regulatory data requirements and attempt to use data-collection opportunities for a variety of purposes. For example, additional data collection can provide valuable insights in developing conceptual models that guide sampling and monitoring

ADDITIONAL INFORMATION ABOUT SAMPLING AND MONITORING AT MINING SITES

"ADTI-MMS Sampling and Monitoring for the Mine Life Cycle" by V.T. McLemore, C.C. Russell, and K.S. Smith in *Securing the Future and 8th ICARD*, conference proceedings, June 23–26, 2009, Skelleftea, Sweden: www.proceedings-stfandicard-2009.com/pdfer/Virginia_McLemore_B4_T3_ADTI-MMS-Sampling-and-Monitoring-for-the-Mine-life-Cycle.pdf

MEND Manual and other reports: www.mend-nedem.org

"Monitoring" in *Global Acid Rock Drainage Guide* (GARD Guide) by the International Network for Acid Prevention: www.gardguide.com/index.php/Chapter_8

"Sampling and Monitoring for Closure" by V.T. McLemore, K.S. Smith, C.C. Russell, and the Sampling and Monitoring Committee of the ADTI-MMS in *Understanding and Responding to Hazardous Substances at Mine Sites in the Western United States*, 2007, edited by J.V. DeGraff. Reviews in Engineering Geology, Vol. 17. Boulder, CO: Geological Society of America. pp. 171–180. doi: 10.1130/2007.4017(11)

USEPA *Abandoned Mine Site Characterization and Cleanup Handbook* (2000), EPA 910-B-00-001: www.epa.gov/superfund/policy/remedy/pdfs/amscch.pdf

USEPA *Framework for Metals Risk Assessment* (2007), EPA 120/R-07/001: www.epa.gov/raf/metalsframework/pdfs/metals-risk-assessment-final.pdf

USGS mine drainage activities: http://mine-drainage.usgs.gov/

USGS Mineral Resources Data System (MRDS): http://tin.er.usgs.gov/mrds/

USGS *National Field Manual for the Collection of Water-Quality Data* (variously dated): http://water.usgs.gov/owq/FieldManual/

USGS watershed contamination from hard rock mining: http://toxics.usgs.gov/regional/mining/

activities. Although regulatory guidelines require specific sampling and monitoring data, broader data collection can ultimately aid in making more reliable decisions.

1.5 SAMPLING AND MONITORING FOR CLOSURE

This handbook endorses the concept of sampling and monitoring for closure during all phases of the mine life cycle on a site-specific basis, from exploration (Hartman 1992) to mine planning, actual mining (operations), reclamation, and closure. Designing for ultimate closure of new mines or mine expansions is a relatively new concept. Because water-quality impacts are a major environmental concern at many mines, detailed assessment of potential impacts on water quality, as well as impacts on the environment in general, are required in most countries before mining can commence. These assessments entail numerous sampling and monitoring programs that begin during the exploration phase and continue throughout the mine life cycle. Geological and environmental variations between mines have fostered a broad range of techniques to understand, sample, monitor, predict, model, mitigate, and control MIW. It is important to ensure that the samples and data quality collected during the exploration phase also are acceptable for subsequent environmental modeling and prediction studies, and that the data can be used for planning for remediation and mine closure.

Mining sites generally are very complex with a variety of specialized sampling and monitoring requirements. The lifetime of a mine extends from the exploration phase to closure and postclosure, and can involve a time frame of many years. Over the lifetime of a mine, a variety of media types are sampled, companies and personnel can come and go, and regulatory requirements can change. These factors make the planning and maintenance of a sampling and monitoring program very challenging. Chapter 2 describes sampling and monitoring needs during the various phases of mining, and the Timing of Environmentally Related Sampling sidebar discusses some considerations in timing of data collection at mining sites.

1.6 TERMINOLOGY

A *sample* is a representative portion, subset, or fraction of a body of material representing a defined population (Wellmer 1998; Davis 1998; Schreuder et al. 2004; Neuendorf et al. 2005; Downing 2010). *Monitoring* is a widely used term, but its meaning can be broad and vague. In this handbook, monitoring is referred to as sampling or collecting data (e.g., visual observations or continuous field readings) over a period of time. Monitoring activities usually follow some sort of structured sampling schedule. All too frequently, sampling is an underappreciated function. It is important to remember that the success of a sampling program depends on the quality, integrity, and representativeness of the samples.

Problematic mine or natural drainage has been given many names including acid rock drainage (ARD), AMD, acidic drainage, and MIW. A discussion of each of these terms is provided in volume 1 of this Management Technologies for Metal Mining Influenced Water series, *Basics of Metal Mining Influenced Water* (McLemore 2008). The distinction between ARD and AMD sometimes has been misapplied, depending on whether drainage quality has been degraded by mining or is due, in part, to natural causes. In practice, the term *ARD* has been applied to all acidic drainage, mining-related or otherwise, and has generally replaced *AMD* in the metal mining drainage literature. Both terms refer to acidic, high mineral acidity, high sulfate (SO_4) waters from mines and natural sources where iron sulfide minerals have weathered in the presence of oxygen and water to form acidic or high metal concentration drainage. But not all adverse drainage from metal mines is acidic; some neutral pH, high metal concentration waters can be detrimental to the environment if not properly managed. In volume 1 of this series (McLemore

TIMING OF ENVIRONMENTALLY RELATED SAMPLING IN THE MINE LIFE CYCLE

Ron L. Schmiermund, Economic and Environmental Geochemistry, Inc.

Timing of environmentally related sampling in the life cycle of a mine is singularly critical to a successful program, but it is an aspect of planning often ignored. Most obvious are baseline-monitoring programs that must take place in advance of planned disturbances. Failure to intelligently implement these programs very early in the exploration/prefeasibility process can result in a poor basis for permit applications and long-term monitoring.

Less obviously dependent on timing are waste rock characterization programs that require reliable drillhole information and can benefit greatly from geologic expertise. The expertise and acquisition of drillhole data co-exist most commonly during exploration and early feasibility phases. In addition, geo-environmental models (GEMs) are most easily and effectively applied and adapted at this time for the purposes of anticipating important variables. Exploration geologists, while available, will understand the implications of their site-specific findings with respect to the basic GEM, and may be able to adapt the drilling plan to provide additional environmentally relevant data. Timing and proactive intercession are critical.

As time passes, all samples deteriorate to varying degrees, but especially in highly sulfidic zones. Sometimes chemical changes and resulting losses of critical information occur in a period of hours or days. In addition, insightful information from project geologists is lost.

Also sensitive to timing is any interaction of environmental specialists with metallurgical testing programs. Such testing typically occurs at multiple times during the feasibility and developmental process, resulting in different and more relevant ore stockpiles and waste streams, each requiring characterization. Sampling of metallurgical process waste streams is best done promptly by persons with a vested interest in the required environmental testing so that relevant particulars of the often complex testing programs are not lost.

Timely interjection of environmental information into an emerging mine plan is the key to having a positive and cost-effective impact on the planning process. If a mine plan is already extensively developed, changes to address potential environmental impacts are likely to receive a cool reception or may simply be impossible to implement. Similarly, a late-stage discovery of significant environmental risks that must be avoided (e.g., leachable metals in critical construction material or abundant sulfides in highwall waste rock) can be very expensive to manage.

2008), the definition of MIW was expanded to include all waters affected by mining and metallurgical processing, which includes wastes from historical operations. This term resolves much of the confusion that exists from using AMD for cases where drainage is from mines but is not acidic, and MIW is the preferred term used to refer to all mining-related waters because acidic, neutral, and alkaline waters can all transport metals, anions, and other contaminants that can potentially impact the environment.

A significant miscommunication problem between scientists, engineers, regulators, and stakeholders (e.g., the public) is that there can be different definitions and interpretations for the same word. The legal definition can differ somewhat from the scientific definition or the common usage by mining companies, scientists of different disciplines, or the stakeholders. It is important that everyone speak the same language and use the same definitions. An example of one of these confusing terms is the word *clay*, which can refer to either particle size or a group of minerals (see Clay Size and Clay Minerals sidebar). Common scientific definitions are used wherever possible in this handbook. The introductory volume of this series (McLemore 2008) includes a comprehensive glossary that defines numerous technical terms relevant to and consistent with this handbook. It also introduces many concepts pertinent to mining.

CLAY SIZE AND CLAY MINERALS

Virginia T. McLemore, New Mexico Bureau of Geology and Mineral Resources

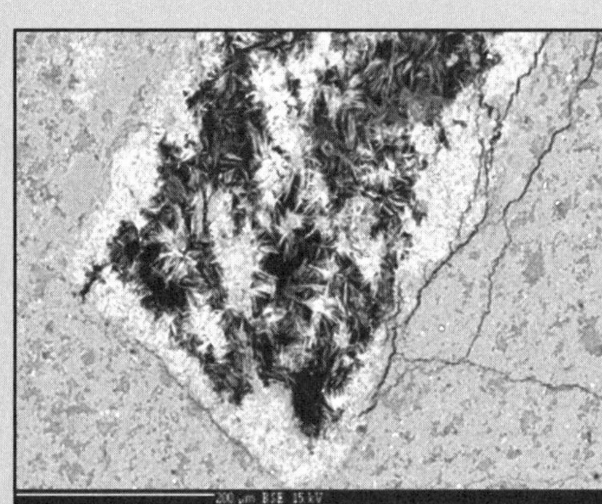

Photomicrograph of clay minerals (mostly sericite) replacing feldspar phenocryst in QSP (quartz-sericite-pyrite)-altered rhyolite (Amalia Tuff) from the Questa molybdenum mine, New Mexico. This sample is an unweathered drill core sample of the ore body in situ before mining.

Clay can be a confusing term that describes a specific group of minerals as well as a specific range in grain or particle size. Grain size distribution is an important physical characteristic of a soil and is different from the quantity of clay minerals. The gradation curve and the percentage of fines control the shear strength and compressibility of a soil or mine-rock pile. Clay minerals are hydrous aluminum phyllosilicates, typically with variable amounts of iron, magnesium, alkali elements, and other elements. Clay minerals are fundamentally built of tetrahedral and octahedral sheets. Geologists and soil scientists usually consider clay-size material to occur at a particle size of <2 µm (clays being finer than silts). Sedimentologists typically use <4 to 5 µm, and colloid chemists use <1 µm to define clay materials. Geotechnical engineers distinguish between silts and clays based on the plasticity properties of the soil, as measured by the Atterberg limits of the soil. Hydrothermal clay minerals can actually replace primary and secondary minerals within the rock fragments comprising soils or rock-pile material, and these replacement clays are not reflected in the grain size distribution and, therefore, do not affect the friction angle obtained from direct shear tests. The majority of the clay minerals are as replacements of other sand size minerals (as illustrated in the photomicrograph). Therefore, the total amount of clay minerals is quite different from the total clay-size material, and this clay would not be expected to affect shear strength now or in the future.

GLOSSARIES RELATED TO MINING TERMINOLOGY

Basics of Metal Mining Influenced Water by V.T. McLemore, ed. 2008. Management Technologies for Metal Mining Influenced Water series, Vol. 1. Littleton, CO: SME.

Global Acid Rock Drainage Guide (GARD Guide) by the International Network for Acid Prevention (extensive online glossary): www.gardguide.com/index.php/Glossary

The Northern Miner's glossary of mining and mineral processing terms: www.northernminer.com/resources/tools/glossary.aspx

USEPA's Abandoned Mine Lands Program (online glossary): www.epa.gov/aml/gloss.htm

USGS's glossaries of scientific terms: http://water.usgs.gov/glossaries.html

1.7 REFERENCES

Davis, B. 1998. What is a sample? What does it represent? *Aust. Inst. Geosci. Bull.* 22:39–34.

Downing, B. 2010. ARD sampling and sample preparation. http://technology.infomine.com/enviromine/ard/sampling/intro.html. Accessed 8/23/13.

Hartman, H.L. 1992. Introduction to mining. In *Mining Engineering Handbook*. Edited by H.L. Hartman. Littleton, CO: SME. pp. 3–42.

INAP (International Network for Acid Prevention). 2009. *Global Acid Rock Drainage Guide* (GARD Guide). www.gardguide.com/. Accessed 8/23/13.

Kleinmann, R.L.P. ed. 2000. *Prediction of Water Quality at Surface Coal Mines*. Morgantown, WV: West Virginia University, National Mine Land Reclamation Center. www.techtransfer.osmre.gov/nttmainsite/Library/hbmanual/predictH2O/predictH2O.pdf. Accessed 8/23/13.

McLemore, V.T. ed. 2008. *Basics of Metal Mining Influenced Water*. Management Technologies for Metal Mining Influenced Water series, Vol. 1. Littleton, CO: SME.

Neuendorf, K.K.E., Mehl, J.P., Jr., and Jackson, J.A. (eds.) 2005. *Glossary of Geology*, 5th ed. Alexandria, VA: American Geological Institute.

Popek, E.P. 2003. *Sampling and Analysis of Environmental Chemical Pollutants: A Complete Guide*. New York: Academic Press.

Schmiermund, R.L., and Drozd, M.A., eds. 1997. Acid mine drainage and other mining influenced waters (MIW). In *Mining Environmental Handbook*. Edited by J.J. Marcus. London: Imperial College Press. pp. 599–617.

Schreuder, H.T., Ernst, R., and Ramirez-Maldonado, H. 2004. *Statistical Techniques for Sampling and Monitoring Natural Resources*. General Technical Report-GTR-126. Fort Collins, CO: U.S. Department of Agriculture, Rocky Mountain Research Station. www.fs.fed.us/rm/pubs/rmrs_gtr126.pdf. Accessed 8/23/13.

Skousen, J., Rose, A., Geidel, G., Foreman, J., Evans, R., Hellier, W., and members of the Avoidance and Remediation Working Group of the Acid Drainage Technology Initiative. 1998. *A Handbook of Technologies for Avoidance and Remediation of Acid Mine Drainage*. Morgantown, WV: West Virginia University, National Mine Land Reclamation Center, http://anr.ext.wvu.edu/r/download/45301. Accessed 8/23/13.

TAEM and SENES (Terrestrial and Aquatic Environmental Managers, Ltd., and SENES Consultants Ltd.). 1997. *Guideline Document for Monitoring Acid Mine Drainage*. MEND Project 4.5.4. Ottawa, ON: Canada Centre for Mineral and Energy Technology.

Tremblay, G.A., and Hogan, C.M., eds. 2001a. *MEND Manual: Volume 2—Sampling and Analysis*. MEND Document 5.4.2b. Ottawa, ON: Natural Resources Canada. www.mend-nedem.org/reports/files/5.4.2b.pdf. Accessed 8/23/13.

Tremblay, G.A., and Hogan, C.M., eds. 2001b. *MEND Manual: Volume 6—Monitoring*. MEND document 5.4.2f. Ottawa, ON: Natural Resources Canada. www.mend-nedem.org/reports/files/5.4.2f.pdf. Accessed 8/23/13.

USBR (U.S. Bureau of Reclamation). 2001. *Water Measurement Manual–A Water Resources Technical Publication*. Washington, DC: U.S. Government Printing Office. www.usbr.gov/pmts/hydraulics_lab/pubs/manuals/WMM_3rd_2001.pdf. Accessed 8/23/13.

Wellmer, F.W. 1998. *Statistical Evaluations in Exploration for Mineral Deposits*. Berlin: Springer.

CHAPTER 2

Sampling and Monitoring During the Mining Phases

Virginia T. McLemore, Kathleen S. Smith, and Carol C. Russell

2.1 INTRODUCTION

The mine life cycle can be divided into four main stages: exploration, mine development, operations, and closure/postclosure (Figure 2.1). The introduction handbook to this series (McLemore 2008) briefly describes the mine life cycle; and the following sections briefly explain each stage of the mine life cycle and its relationship to sampling and monitoring for closure.

Figure 2.1 also summarizes the data and information needs during the mine life cycle. During the premining and exploration phases of mining, it is beneficial to address climatological data, a water balance (also called water budget), and habitat disruption. Additionally, samples could be collected to determine potential for contaminant release (including, but not limited to, acid generation), and the receiving environment and receptors could be characterized to determine baseline conditions and possible contaminant pathways. It is advisable that geological units likely to be exposed during mining be fully characterized and evaluated for potential contaminant release.

2.2 EXPLORATION (PREMINING/UNDISTURBED)

The first stage of mining, the prospecting or exploration stage, can best be described as the search for a mineral deposit (Hartman 1992). A *mineral deposit* is any occurrence of a valuable commodity or mineral that is of sufficient size and grade (content) for potential economic development under past, present, or future favorable conditions. An *ore deposit* is a well-defined mineral deposit that has been tested and found to be of sufficient size, grade, and accessibility to be extracted and processed at a profit over a specific time. Most exploration programs do not evolve into an operating mine because the deposit size is too small, the grade is too low, or the environmental risks are too high to be profitable at a given time. Therefore, the economics include an assessment of the quality of the mineral deposit and potential environmental costs, which will dictate the need for sampling and monitoring at this stage (see Figure 2.2).

Mineral exploration can typically require 10 years or more of study before a decision is made to develop or drop a potential mineral deposit. Several feasibility studies can be conducted to determine if the deposit can be mined economically at a specific time. Companies constantly reevaluate formerly discovered projects as economics change with time. Exploration can include literature searches; geological mapping; drilling; geochemical characterization of rock, soil, water, or biological media; geophysical and remote sensing surveys and aerial photography; among other studies (most of these techniques are briefly described in Appendix 3). A tremendous amount of

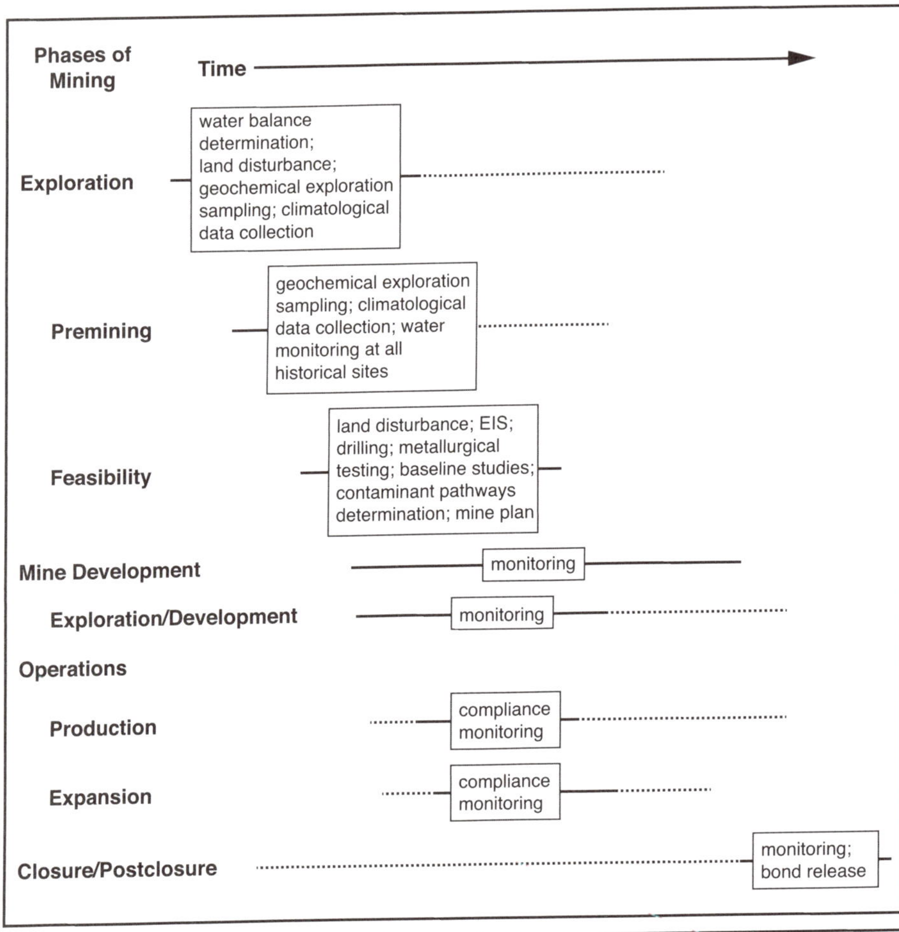

FIGURE 2.1 Data and information needs during the various phases of mining. Operations can include advance exploration and continuous planning for closure. Closure can occur at any stage.

data collected during exploration can be useful later in identifying, predicting, and mitigating mining influenced water (MIW).

Exploration has the potential to disturb the baseline conditions, and typically these disturbed areas need to be reclaimed. The environmental impacts of exploration can range from habitat disruption to surface-water and groundwater impacts from improper management of fuel spills. Exploration commonly requires trenching or drilling to retrieve rock samples for assay or other chemical analyses or metallurgical testing. Exploration roads, drill pads, and trenches have the potential to damage water and drainage quality from the erosion of sediment alone. These trenches and drill sites and their access roads should be sampled prior to disturbance and reclaimed after exploration, unless further development of the deposit is planned. Geochemical sampling, road construction, and drilling, which typically occur during the exploration stage, can continue into the mine development stage and throughout the mine life cycle.

Acidic drainage (AD) can result where outcrops of sulfide-bearing rock are disturbed. One common misconception is that ores near the surface contain no sulfide minerals because they have been oxidized or weathered. Whether sulfide minerals exist at the surface depends on local climate and its effects on factors such as

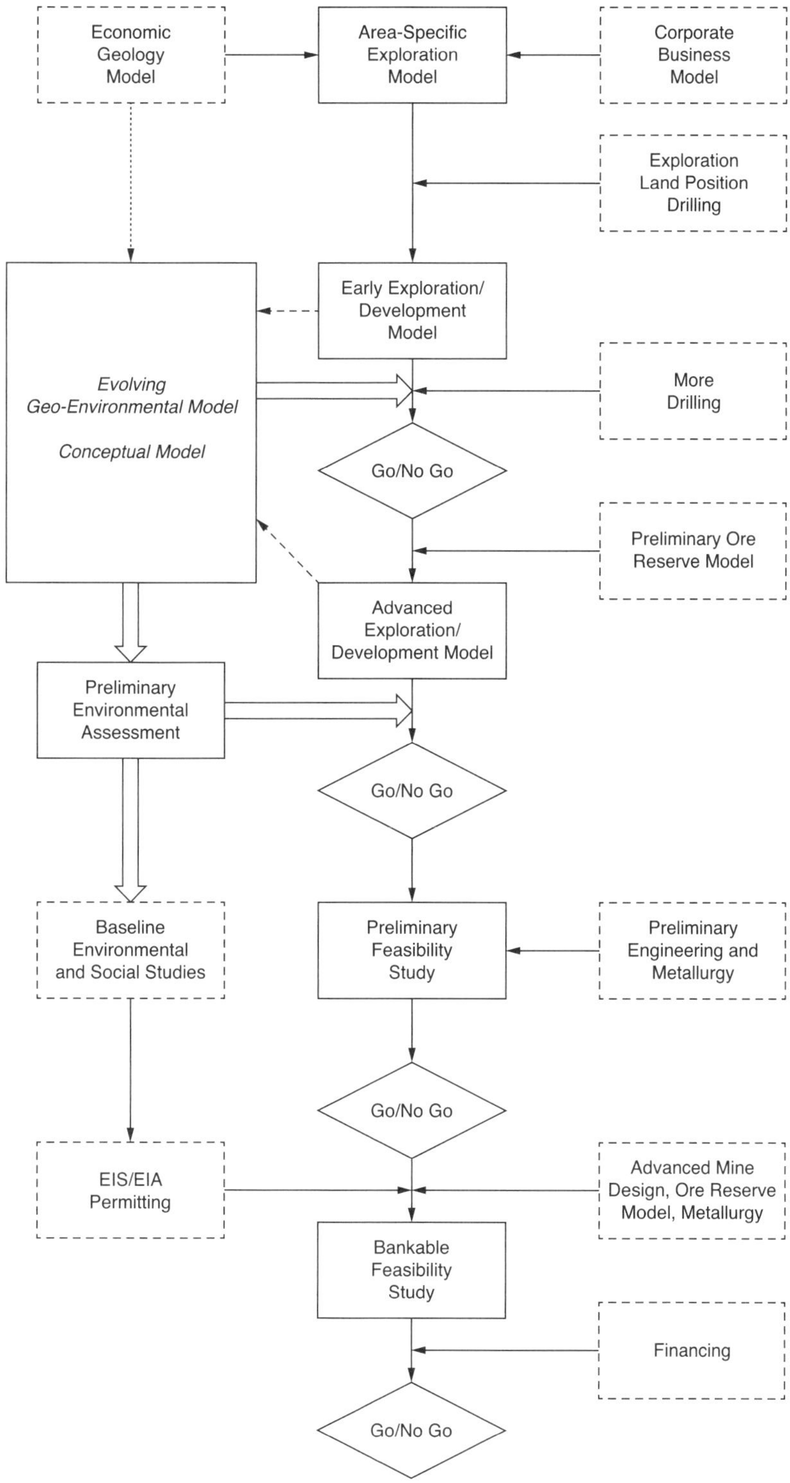

Source: Adapted from Schmiermund and Ranville 2006.

FIGURE 2.2 Flowchart summarizing the major steps during project development and integration of geologic, social, economic, and environmental inputs in feasibility studies to determine if a project can proceed to the next stage in the mine life cycle

ACID DRAINAGE FOUND DURING CONSTRUCTION OF TAILINGS DAM AT THE KENSINGTON MINE IN ALASKA RESULTS IN ADDITIONAL MONITORING

Virginia T. McLemore, New Mexico Bureau of Geology and Mineral Resources

The Kensington gold mine is approximately 72 km (45 miles) north-northwest of Juneau, Alaska, and began production in June 2010 by Coeur Alaska. The gold is hosted by multiple mesothermal, quartz-carbonate-pyrite swarms and veins within the Cretaceous Jualin diorite. The mine is expected to produce 108,000 to 114,000 ounces of gold in 2013 at a grade of 0.27 ounces/short ton using underground mining techniques. Kensington's proven and probable reserves were 1.0 million contained ounces of gold at year-end 2012 (Coeur Mining, Inc., n.d.). The Kensington deposit is in carbonate rocks, and, therefore, the waste rock piles and tailings piles are not net-acid generating, because the carbonate rock will neutralize any acid that is produced. However, during construction of the tailings dam, pyrite-bearing graphitic phyllite was exposed to air and generated acid drainage near the site (Golden 2008; Alaska Department of Environmental Conservation 2009). Modifications to the existing reclamation plan were required. Reclamation of the acid drainage was followed by plans requiring special handling of any excavated pyrite-bearing rock and rigorous sampling and monitoring plans. This example illustrates

- The importance of characterizing all geologic materials that will be encountered during construction;
- That despite all geologic characterization studies and mine plans, unexpected problems can occur during the construction and operation stages that could require changes to the original mine plan, including additional sampling and monitoring; and
- The quick response required to remediate the problem.

- Weathering rates,
- Mineral type and composition,
- Length of exposure to oxygen (time since uplift or exposure),
- Location of rocks above the local water table, and
- Quantity and rate of water flow through the rocks.

Generally, higher accessibility of sulfide minerals to moisture and oxygen promotes oxidation, which is why sulfide-bearing rocks tend to weather more slowly in dry climates compared with wet climates. Sulfides are not present in all mineral deposits, but the possibility of acid rock drainage (ARD) should be considered during both shallow and subsurface exploration. Any material with sulfide minerals should be sampled. Furthermore, country rocks that could be affected during construction of mine facilities should be sampled and characterized to determine if sulfides are present. For example, during construction of the tailings dam at the Kensington mine in Alaska, pyrite-bearing graphitic phyllite was exposed that generated ARD, even though the Kensington deposit is in carbonate rocks and therefore non-acid generating (see Acid Drainage sidebar).

In the past, prospectors typically panned in streams and sampled at the surface to identify potential areas for continued exploration. Once a likely area was identified, the prospector would typically dig pits, shafts, or adits to test the mineral occurrence and determine the grade and extent of the deposit. Commonly, the prospector proceeded directly into mining the deposit as long as minerals were visible. Many such formerly discovered prospect sites have AD because pyrite, the nearly ubiquitous indicator of most types of metallic ore bodies, was disturbed at the surface and broken into chunks that exposed more pyrite to oxygen and water. This exposed pyrite then weathers to form AD.

Traditional and modern exploration technologies, such as satellite and other remote-sensing imagery, geologic maps, geochemical and geophysical studies, and geologic reports, can be used in environmental studies (see summaries of these techniques in Appendix 3). Traditional exploration samples from stripping, drilling, trenching, bulk sampling, and even panning can provide useful environmental insights into potential water- and air-quality problems. Extra chemical analyses of sediment and soil samples and biological materials collected during the exploration stage can be used for establishing baseline environmental monitoring data. Geophysical techniques, which explorationists use routinely to predict the depth and breadth of mineral deposits, can be used to characterize the probable extent of natural AD sources prior to mining (see summaries of various techniques in Appendix 3). Geologic maps, the foundation of mineral exploration and mining, can provide fundamental information useful in interpreting baseline environmental implications of geochemical analyses.

Environmental evaluations in mining are useful not just for site characterization, but for determining if a deposit is economically feasible. The economic feasibility determination must consider baseline or ambient conditions and the costs of closure. Where mineral deposits associated with sulfide minerals occur near the surface, it is likely that current soil productivity, natural drainage quality, and biological conditions are affected or impaired by minerals associated with that deposit. Furthermore, these unique environmental conditions can support rare or endangered species that have colonized these unusual habitats. Defining these conditions before mining begins can be critical in evaluating the economic and environmental liabilities of mine development. See the Objectives of Feasibility Studies sidebar.

Another economic assessment consideration includes the proximity to wilderness areas, national parks, wildlife refuges, recreation sites, or other valuable or fragile environments. The presence of threatened or endangered species, or their habitats, can make or break a project. Knowing these features in advance of or during exploration commonly determines the "go or no-go" decision. Some of these features should be sampled if the project moves forward. A significant gold deposit in Montana near the border of Yellowstone National Park drew unprecedented opposition from citizens concerned about how mining might affect the park. Their opposition was perhaps of key importance in the company's decision to abandon its plans for this deposit (Russell 2006; see New World Mining sidebar). Similar public opposition to the Pebble mine in Alaska, which is located in a major salmon habitat, led Anglo American to pull out as a partner in the venture.

Regulations governing exploration programs vary from jurisdiction to jurisdiction. Some exploration projects are so extensive that operational and reclamation plans must be developed prior to drilling. Permitting is required for exploration in most areas in the United States and elsewhere, and reclamation of disturbances caused by exploration is also required in most areas.

2.2.1 Sampling and Monitoring During the Exploration Stage

Information and data collected during exploration can be applied to later stages of the mine life cycle. By slightly expanding data collection during this stage, exploration sampling and monitoring becomes essential for the foundation of later mining stages, especially mine planning, development, and closure. As the first to set foot on potential mine sites, exploration geologists

OBJECTIVES OF PREFEASIBILITY AND FEASIBILITY STUDIES

Virginia T. McLemore, New Mexico Bureau of Geology and Mineral Resources

A *mine prefeasibility study* or a *preliminary feasibility study* is a preliminary comprehensive study of the viability of a mining project where a mining method and effective method of mineral processing has been established (CIM 2005). A mine feasibility study is a more detailed evaluation of a mine project to determine if that project has the likelihood of becoming an economic, operating mine. Feasibility studies are generally performed during exploration and mine development but can also be performed during any stage of the mine life cycle. Besides determining the location, accessibility, geology, climate, topography, resources and reserves, processing, transportation, market and price expectations, and infrastructure, various risks should be identified and evaluated, including if the property is under company control or outside company control, is the community generally favorable for the mining operation, what are the environmental risks, political climate, and so on. Guidelines for prefeasibility and feasibility studies can be found in the Canadian National Instrument (NI) 43-101 requirements or the Australasian Code (JORC 2012).

Different types of feasibility studies are required during the exploration stage to determine if the project could be an economic mine and determine whether more money should be spent on the project. Typically, several feasibility studies are required and can have different names, such as scoping studies (order of magnitude), prefeasibility studies, final feasibility study, detailed feasibility study, and so forth, depending on company policy. Most exploration projects do not become operating mines, because the feasibility study at a specific time determined that the project would not be economic at that time. Additional information, change in company, or change in economic conditions can result in a decision change and new feasibility studies are generally required. The accuracy of each feasibility study improves as the project becomes closer to becoming an operating mine (see figure), because more knowledge is obtained as the company spends more money. The early feasibility studies provide estimates of the likelihood of the property becoming an operating mine, and later feasibility studies will be more accurate. The results of the feasibility studies should be incorporated into the final mine plans.

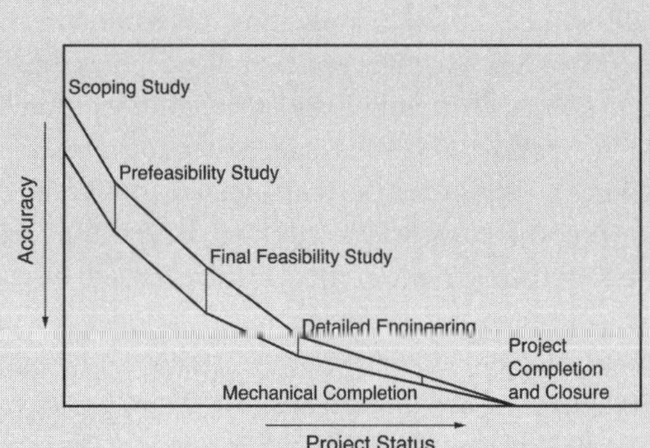

Source: Adapted from McCarthy 2010.

Probable accuracy of each phase during an exploration project

have a chance to develop insights and fill many future data needs, provided they sample, analyze, and preserve data and information with a secondary objective toward the later mining stages and closure.

Explorationists analyze more ground than is mined and can produce a regional geochemical and environmental picture that can be applied to environmental planning and monitoring. For example, by noting areas prone to flooding, collapse, or landslides, explorationists can lead mine planners away from such dangerous or unstable features. Recent studies near Lake City, Colorado, show that head scarps of major landslides lie in hydrothermally altered rock that is unstable

NEW WORLD MINING HALTED BECAUSE OF PROXIMITY TO YELLOWSTONE NATIONAL PARK, MONTANA

Carol C. Russell, U.S. Environmental Protection Agency

The New World mine was proposed in 1990 in a previously mined area near Cooke City, Montana, just northeast of Yellowstone National Park. The controversy took on national and worldwide significance when in 1995, a delegation from the World Heritage Commission visited the park to investigate the proposed mine and placed Yellowstone on the list of World Heritage Sites in Danger. A *New York Times* reporter stated that "this [author] page is convinced that the proposed New World mine is a disaster-in-waiting that could ruin one of America's leading ecosystems" (Semple 1995). The mine plan was good and included procedures to clean up all of the abandoned mine wastes in the area. However, the end result was the purchase of the mine by the U.S. government with the U.S. Forest Service (USFS) taking on the responsibility of acid rock drainage and remediation of the previously mined areas. Remediation by the USFS is ongoing. See USFS Web site (www.maximtechnologies.com/newworld/index.htm) for more information.

because of rapid weathering caused by pyrite oxidation (Diehl and Schuster 1996). This alteration is related to nearby mineral deposits of gold and sulfide veins. Similar landslide scarps are found in other areas of extreme alteration, such as at the Questa molybdenum mine in New Mexico (Figure 2.3; Vincent 2008; McLemore 2009) and the upper Alamosa River Basin in Colorado (Bove 1994). Exploration work integrated with soil chemistry and physiographic analyses could be used to identify unstable areas to avoid during construction of waste rock piles and other mine facilities.

By recording observations of media other than rocks, such as plant life, soil characteristics, and water chemistry, the explorationists lend insights into native environmental conditions, and set the stage for monitoring programs and features that can minimize impacts to the environment. However, cursory visual observation that is incomplete or done by persons who are not properly trained can lead to skewed or incorrect interpretations and should be avoided. Hydrologic features, such as wetlands, seeps, and iron bogs, can indicate where MIW will surface over the mine life cycle. Subsidence and historical mining structures tend to affect how subsequent mines are planned and developed, and, because explorationists tend to discover such features through ordinary ground studies, their observations can be useful for mine planners.

Explorationists tend to concentrate on analyzing potentially valuable elements, such as gold, copper, and zinc, and analyses of *pathfinder elements* to assess large areas and focus on smaller drill targets. Even today, not all exploration geologists will use complete trace element analyses. If, in addition to these limited analyses, a broader suite of chemical parameters are analyzed, data thus produced can be useful for studies as diverse as geochemical modeling, weathering, and slope stability, if the data meet data quality objectives. Complete chemical analyses of water, soil, sediment, rock, and vegetation are commonly necessary to determine the background or baseline geochemistry and biochemistry of a site. If exploration and environmental sample results are integrated, there is a better chance of identifying problems that mining will encounter. For example, at the Kendall mine in Montana, thallium was found to be a problem long after mining had begun (W. McCollough, personal communication, June 2003). Early identification could have allowed incorporation of management practices to mitigate or minimize potential problems. The difference in cost between complete chemical analyses and limited analyses of only pathfinder elements can be insignificant compared with the potential consequences of inadequate documentation of

Courtesy of Virginia T. McLemore.

FIGURE 2.3 Landslides and debris flows in the Questa alteration scar above the Questa molybdenum open pit mine, Taos County, New Mexico

premining conditions. However, it is important to realize that these up-front costs are significant to limited exploration budgets. Also, it is important to keep in mind that analytical limits of detection adequate for determining premining environmental conditions can be lower than those needed for exploration. Whenever possible, it is good practice to retain all data and samples, even if they give negative results for exploration purposes. Properly archived data and samples could be essential during reclamation and closure because they can provide the only premining information. Data needs to be archived and backed up continuously, and samples should be stored such that they can be readily retrieved for additional analysis. Even data from formerly discovered projects can be important in establishing baseline or premining conditions in other areas and aid to a better understanding of geological processes that lead to adverse MIW conditions (see Schmiermund et al. 2006).

Efforts should begin early in the mine life cycle to record climatic conditions in areas of exploration interest, because these data are required for defining a water balance for the mine site and regulatory purposes, including defining baseline and/or background conditions. Parameters of interest include

- Annual and seasonal precipitation rates and intensity,
- Daily temperature range,
- Relative humidity,
- Predominant wind direction,
- Wind speed, and
- Evaporation rates.

Seasonal data are necessary, and supplementary climatological data can be required to adequately characterize the site. Weather stations can seem like a waste of exploration dollars, especially the collection of that data, but the data thus developed serve as a basis for water balance calculations that will be needed throughout the mine life cycle. Moreover, on-site climatic data

CLIMATE DATA AT THE SUMMITVILLE MINE, COLORADO

Carol C. Russell, U.S. Environmental Protection Agency

Summitville is a classic example in the United States where the water balance and precipitation calculations were based on data from the nearest town, Alamosa, 75 km (46.6 miles) away and more than 2,000 m (6,562 ft) lower in elevation. Very cold temperatures were unanticipated when the liner for the cyanide heap leach was laid in place. Failure of the liner occurred almost immediately, releasing acid rock drainage and cyanide (Gray et al. 1994). "The polluted water was supposed to be contained in a heap leach pond, but the state's oversight of the operation was lax. Unusually heavy snow and runoff overwhelmed the pond's critical systems, sending cyanide-laced liquid into tributaries of the Rio Grande, killing all fish downstream for 17 miles [27 km]" (USEPA 2011a). In addition, no one expected to have snow depths of more than 8 m (26 ft) the first winter the government took over the site. The unknown effects of the water balance and weather are factors contributing to the mine's failure. See King (1995) for more information.

can be far more accurate, and thus more applicable for mine design, than data collected from the nearest weather monitoring stations, which can be distant in location or climatic conditions. For example, at Summitville, Colorado, where climatological data came from a site 75 km (47 mi) away, the site-wide hydrologic balance underestimated the precipitation, which led to critical shortcomings in the water balance and the site's water management needs (Pendleton et al. 1995). Many climatic stations can be required at some sites to adequately characterize different microclimates because of elevation and topographic changes. Such is the case at the Questa molybdenum mine, New Mexico, where elevations range from 2,377 to 3,048 m (7,800 to 10,000 ft), and slope aspects vary from north- to south-facing, which places some slopes directly in the sun where evaporation/sublimation is the greatest, while other slopes are in the shade where snowpack is present much later in the season (McLemore 2009). See the Climate Data sidebar.

Surface and subsurface rock samples can provide essential mining and environmental information, and closure/postclosure needs should be considered in the choice of collection methods. Surface-rock samples are collected with either manual tools or a backhoe. Rocks from below the surface can be collected with a variety of drilling methods, including churn, hammer, reverse circulation, and diamond drilling. The former three drilling methods produce rock chips or powders, whereas the diamond-drill method produces a continuous core of the rocks it traverses. Diamond-drill cores provide a more complete record of the original stratigraphy of overburden, waste rock, and mineralized rock than samples from rotary methods, and are considered the most useful for collection of detailed geologic data. Selected exploration drillholes can be converted to monitoring wells to provide samples for baseline, operational, and postclosure conditions or other monitoring needs.

Prefeasibility and feasibility studies are employed at various stages of the exploration project to determine if the project could result in an ore deposit that could be extracted economically at known and projected costs with a profit. Further details of types and uses of prefeasibility and feasibility studies and evaluating risk are presented by Lord et al. (2001), Dowd (2002), Schmiermund and Ranville (2006), and others who have described the process, which is summarized in Figure 2.2. Part of any prefeasibility or feasibility study should be the identification of geologic conditions that could impact surface water and groundwater. One such method is the U.S. Geological Survey's (USGS's) geo-environmental models (GEMs) (Plumlee and Nash 1995); some examples of their use are summarized in Appendix 6. See the GEMs sidebar.

GEO-ENVIRONMENTAL MODELS OR GEMS

Virginia T. McLemore, New Mexico Bureau of Geology and Mineral Resources

Geo-environmental models (GEMs) provide one method of developing site conceptual models and identifying potential water quality impacts based on characteristics of the mineral deposit. A GEM is a compilation of geologic, geochemical, geophysical, hydrologic, and engineering information pertaining to environmental behavior of geologically similar mineral deposits prior to mining and resulting from mining, mineral processing, smelting, and refining (Plumlee and Nash 1995). Geologic data are compiled, including deposit type, related deposit types, deposit size, host rocks, surrounding geologic terrain, wall rock alteration, nature of ore, mining and ore processing methods, deposit trace element geochemistry, primary mineralogy and zonation, secondary mineralogy, soil and sediment signatures, topography, hydrology, drainage signatures, climatic effects, and potential environmental concerns (see Appendix 6; Plumlee and Nash 1995). Thus, sites can be initially assessed by determining site characteristics and comparing them with previously determined relationships between similar characteristics and their potential for adverse impacts on water quality. GEMs have more applications in the early stages of the mine life cycle.

2.3 MINE DEVELOPMENT

Once an economic mineral deposit is discovered, its limits defined, and the feasibility studies indicate that an economic deposit exists, then permits must be obtained so that mining can proceed to the second stage of the mine life cycle, mine development. Baseline environmental conditions characterized during the exploration stage should be extended into the development phase in greater detail and any information gaps addressed during the earliest part of development, prior to ground disturbances. This is because disturbances during development and mining can typically be environmentally indistinguishable from naturally degraded conditions in near-surface mineral deposits. It is important to incorporate appropriate natural conditions into the site plan to set realistic reclamation goals. Permitting, operational, and preliminary mine-closure plans are generally developed at this stage. Mine development of sites typically requires additional infill drilling to refine the three-dimensional features of the deposit to sufficiently develop plans for excavating the deposit. This phase, which is commonly conducted by a contractor, consultant, or a junior mining company, also includes construction of mine access and mine-haul roads, rail lines, and terminals (if they are required); installation of electrical, gas, and other utilities; and construction of mine and process buildings, support structures, and, in some areas, housing facilities. In remote or undeveloped areas, facilities construction can include housing and social infrastructure, such as schools, churches, recreation sites, and shopping centers. Excavation requires heavy machinery and the use of petroleum and other industrial products for operation and maintenance. If the mine site is not pristine prior to exploration, chemical scans can be useful to establish premining baseline conditions. All excavated sites should be examined and characterized to determine if any acid-generating material was encountered and to determine long-term stability of slopes.

Before engineered mine-waste and process-waste repositories, roads, buildings, or other structures are constructed on a mine site, condemnation drilling should occur; that is, closely spaced drillholes are bored to ascertain that viable minerals and historical mine workings do not lie beneath the proposed structure. Information collected through such drilling can be applied to geochemical and hydrological balance studies.

The mine development stage can also include construction of metallurgical processing facilities, such as mills, pipelines, conveyance facilities, engineered tailings impoundments, containment facility liners, leach pads, and solution collection facilities. Metallurgical testing and development are locally conducted on-site at metallurgical pilot plants. For operations requiring cyanide or acid leaching, mills, leach pads, processing ponds, and their related infrastructure, such as cyanide detoxification facilities, are constructed during the development stage, and leach pad construction can continue into the operational stage, especially at large, multiyear operations.

Metallurgical testing provides critical data for evaluation of ore and has environmental applications at mining and processing sites. Metallurgical testing is conducted prior to the mine development stage but is generally at a smaller scale that provides data for preliminary economic assessments or feasibility studies. Depending on the type of processes ultimately selected, metallurgical testing will generate ore composites, leach residues, process solutions, or tailings that can be made available for geochemical testing to evaluate the potential for ore, leach residue, and tailings to generate MIW/ARD in the pit walls and stockpiles, on leach pads, and in tailings storage facilities. Metallurgical testing solutions provide important information on the concentration of metals and anions in process solution, which is particularly important for process facilities with a positive water balance requiring excess water to be treated and discharged. Also during this metallurgical testing, contaminants arising from the solutions and slurries that come from mineral processing should be evaluated to determine if they contain elevated levels of contaminants. For example, some of these slurries could have elevated levels of arsenic that are released during mineral processing. Metallurgical testing produces assays and geochemical analyses similar to many that are run during exploration, such as detailed multi-element analyses. These analyses describe the rocks and minerals at a very detailed mineralogical and physical scale to enable effective recovery of the valuable commodity. During this phase of testing, trace-element composition of minerals is examined, and from the environmental perspective, this testing can be important for understanding of site-wide geochemical characteristics that can be useful in ore and waste rock management. For instance, cadmium is commonly present as a trace element in sphalerite (ZnS), but sphalerite in some deposits is not of sufficient quantity to warrant recovery during milling. In such cases, if waste leaching or weathering testing shows that the material containing sphalerite could cause an environmental impact, the waste storage facility could be engineered so that it could be specially handled by placing it in a specific, more isolated area of the waste rock facility instead of with other gangue minerals (noneconomic minerals, such as quartz or calcite that are part of an ore deposit) to avoid release of cadmium. Additional environmental protections can also be put in place to minimize the distribution of cadmium and zinc into the environment. Such detailed mineralogical observations are typically made during metallurgical testing. See the Why Mineralogy Is Important sidebar.

Material strength properties, which are typically determined during mine development, can provide useful environmental and closure information. Whereas mining engineers examine rocks to determine the maximum height or depth of mine benches, the rock properties can also be used to predict the conditions of potential failure of mine features during operations and after closure. For example, if a mine pit fills with water after mine closure, pit walls can become unstable. Slope failures after mining can be as much a problem for mining engineers as are slope failures during mining. Areas of potential collapse or subsidence can be indicated as well, because these collapses can affect water quality, containment, and water flow. In areas of historical mining, past underground workings that are accessible need to be mapped and characterized for potential stability and drainage issues.

WHY MINERALOGY IS IMPORTANT IN ACID MINE DRAINAGE STUDIES

Stephen J. Sutley, U.S. Geological Survey

A remedial investigation is being undertaken at Holden mine in the northern Cascade Range in north-central Washington under the supervision of the U.S. Forest Service, the U.S. Environmental Protection Agency, and the Washington State Department of Ecology to determine the presence, magnitude, nature, and extent of potential environmental concerns associated with this historical mining endeavor. Between 1938 and 1957, more than 10 million short tons of copper, zinc, silver, and gold ore were extracted from this deposit.

Mineralogical studies, mostly involving X-ray diffraction techniques, have proven critical in the characterization and interpretation of prevailing site conditions. Some of the major issues addressed and resolved by mineralogical studies include (1) verification of the primary sulfide minerals (pyrite, pyrrhotite, chalcopyrite, etc.) ultimately responsible for the site's acidic, metalliferous discharge; (2) identification of various soluble secondary sulfate minerals (including iron, copper, and zinc phases) known to have a major impact on the seasonal variations in the site's water chemistry (as a result of the soluble sulfates cyclic dissolution and hydraulic flushing); and (3) determination of the mineral residences for metals detected in the site's surface and subsurface soils (permitting an estimation of the bioavailability of metal toxicants).

These, and other studies, will enable government agencies, consultants, and other concerned groups to evaluate the effectiveness of proposed remedial measures and determine the most practicable treatment alternatives as they apply to the Holden mine cleanup.

2.3.1 Sampling and Monitoring During the Mine Development Stage

Mine development is the transition stage between exploration and mine operations. Hence, many issues considered during exploration also need to be addressed during this stage of mining. If collection of on-site climatological data did not begin during exploration, it is essential that this information be collected early in the mine development stage, even if the local regulation does not require them. In order to establish successful revegetation at the time of mine closure, observations of native vegetation, soil depth, slope and aspect, and other preexisting conditions must be documented. This information is also useful during the operations stage for planning placement of waste materials; it is more economical to move waste only once. During the early stages of development, it is important to consider reclamation. An example of this practice is when Nevada Placer Dome hired the Duckwater/Shoshone tribe to collect seeds and grow plants needed for revegetation (B. Upton, personal communication, September 2003).

During the mine development stage, metallurgists evaluate detailed mineralogy. Concurrently, others are developing a water balance for the site and defining baseline environmental conditions. It can be extremely useful at this stage to set aside an area for site weathering tests and initiate predictive tests, such as acid–base accounting, meteoric water mobility, static and kinetic tests, and other long-term weathering tests, conducted under natural local conditions. Some of these tests could begin earlier in the mine life cycle. These data are useful in the prediction of acid-generation and water management requirements at the site. For example, the acid-generation characteristics of the waste rock at the Zortman–Landusky mine in Montana changed over time (Miller and Hertel 1997) because of the buildup of ferric iron, a strong oxidant, in the sulfidic waste-rock piles and the heap pile. This unanticipated change in mineralogy contributed to the eventual bankruptcy of the parent company (Russell 2006). Air monitoring can also begin during this stage, or earlier, to develop baseline conditions prior to ground disturbance.

Permitting, which generally takes place during the mine development stage, can require significant environmental data to accommodate the permitting process. The mine development stage is a good time to design and implement compliance-monitoring programs. Lysimeters can be installed at this time to measure depth to water and to aid in evaluating infiltration rates. Pilot studies, such as test heap-leach pads and site waste-rock weathering columns or test pads can be constructed at this time, and monitoring can begin. It is good practice to continue these site weathering and closure pilot studies throughout the remainder of the mine life cycle to provide information for closure. For example, when research on pilot heap closures was undertaken at the Golden Sunlight mine in Montana with the U.S. Environmental Protection Agency's (USEPA's) Office of Research and Development and MSE Technology Applications, Inc., the operator (Placer Dome) found pilot-testing results to be very useful and is now undertaking similar tests at their operations in Nevada (B. Upton, personal communication, September 2003).

Transport and pathways of waters and potential contaminants need to be identified during this stage. Detailed geological maps, including structural information, should be refined and updated. Potential biological receptors can be evaluated and indicator organisms established.

Disturbance from road building and other structures should be sampled and characterized for use during closure. It is important to keep in mind that closure can occur any time during mine development if feasibility studies determine that continued development or mining cannot be economical. If closure occurs, all disturbed ground will have to be reclaimed.

2.4 OPERATIONS

The operations stage of mining includes the process whereby ores and waste rock are removed from the ground, milled and/or leached, and concentrated into a salable product, such as a metal sulfide concentrate for smelting or high-grade gold or silver bullion for refining. Sampling of the ore and waste materials is necessary in order to know what potential contaminants exist in the rock mass, and monitoring of groundwater or surface water can be necessary for environmental compliance. Air pollution monitoring also could be required. Excess water produced by the operations may require sampling during treatment and discharge. Much of this excess water is reused elsewhere in the mine and process operations.

2.4.1 Excavation

The initial and usually ongoing phase of the operations stage is excavation, which includes the extraction of ores and waste rock by drilling, blasting, and mucking (removing the broken material from the mine). During excavation, as rock is removed from the ground, excess water may need to be pumped or drained from the area of underground and surface workings in order to accommodate miners and equipment; sampling of this water is important for closure. At open pit mines requiring dewatering, water is typically pumped from wells adjacent to the pit, a process that lowers the water table below the bottom of the pit, and from sumps within the pit, if necessary. In underground mines, excess water is usually pumped from sumps that are constructed in the lower reaches of the excavation and piped upward through vertical mine shafts to the surface. At many older underground mines located in steep terrain, such as in Colorado, underground mining was made possible through construction of an adit (horizontal shaft) or tunnel that drained groundwater from one or more mines to a nearby stream. Such drainage tunnels, which were constructed several kilometers (miles) underground, could dewater many square kilometers (miles) of surrounding rock and workings, and thereby reduced the need for expensive pumping of water from mine workings. However, drainage-tunnel water is commonly of poor quality and could require treatment. Today, much of this water is recycled and used in operations.

Whenever the water table is lowered to allow mining below it or when it fluctuates (such as occurs seasonally or during storm events), there is a chance that water quality will degrade because of oxidation of sulfide minerals. When this degradation happens, acid water or any MIW can form wherever there are sulfide minerals that had resided below the water table. Monitoring these waters during mining, both inside and outside the mine permit area, will be important in planning for mine closure. Dewatering that occurs during the operation stage may have to continue after excavation ceases in order to control environmental impacts. If a mine floods, resulting MIW can require treatment or other special management. See volume 3 of this series (Castendyk and Eary 2009) for more details on chemistry and dynamics of pit lakes at mine sites.

2.4.2 Milling, Concentrating, and Leaching

After ores are excavated from the ground, they are generally placed in engineered rock piles (i.e., dumps) or reduced to a nominal size to allow ore minerals or leachable metals to be segregated from waste rock or gangue minerals or to promote rapid leaching. Crushing and milling generate a rock mass with a surface area that is many times greater than the original rock mass. Once exposed to air, finer-grained crushed and milled rock can generate acid if it contains acid-generating sulfide minerals, such as pyrite. Sulfide ores are typically milled to a fine-grain size (generally <100 μm, although this varies). This allows particles of ore minerals to be segregated from gangue minerals, or from other ore minerals, via flotation, gravity, electromagnetic methods, or other processing means, or they may be directly leached. Other types of ores, including many gold and copper-oxide ores, do not require further size reduction (dump leaching), or they need only be reduced to larger chunks (typically 2.5–12.7 cm [1–5 in.]) so they can be heap leached after being placed in a lined facility and treated with a leaching solution to remove the economic metals. Dump-leach materials are processed with no further size reduction. Size reduction is also used to promote rapid leaching. Some gold ores are milled to a fine-grain size to expose gold-bearing minerals to leaching solutions or roasted/autoclaved to oxidize the carbon or sulfidic minerals. Tailings are finely ground waste from milling followed by metal extraction and are managed by dewatering and storage, primarily in lined facilities.

Mineral separation techniques have changed over time, and the sampler must be aware of historical and modern technologies, because different technologies can lead to potentially different environmental consequences (W.R. Jones 2007). For example, stamp mills were commonly used at gold mines during the late 1800s and early 1900s. These mills ran gold ore through a number of levels, with cam-driven hammers pounding the ore into finer and finer particles. Gold was separated from the finest particles using mercury amalgam. Waste materials from stamp mills commonly contain fine-grained sulfide minerals that can result in acidic metal-rich drainage, and mercury contamination can be found in the vicinity of some historical stamp mills.

Leaching is a common technique in which valuable, leachable metals are removed from the gangue, typically along with other less desirable metals. The metals are then extracted from the high-metal concentration pregnant solution by other hydrometallurgical processes (such as electrowinning, zinc cementation, and carbon adsorption). Leaching chemicals, such as cyanide and sulfuric acid, can be potential environmental contaminants. See Uravan Mill sidebar.

2.4.3 Smelting and Refining

Although not always the case, metal smelters and refineries today are large, specialized facilities that are not profitable enough to operate at most individual mine sites. Consequently, many are located in areas away from mining districts and service multiple mines; many of these areas are in

URAVAN MILL, COLORADO

Carol C. Russell, U.S. Environmental Protection Agency

The 1.8-km^2 (450-acre) Uravan uranium mill began as a radium-recovery plant in 1915. Its owners later converted it for vanadium extraction. From the 1940s to 1984, the plant utilized sulfuric acid in the uranium and vanadium processing facility, thus the origin of the town's name, Uravan. Operations at the site left a large volume of acidic and alkaline wastes, contaminating air, soil, and groundwater near the plant and the San Miguel River. Contaminants included radioactive products such as raffinates, raffinate crystals, and mill tailings containing uranium and radium. Other chemicals in the tailings and groundwater were heavy metals such as lead, arsenic, cadmium, and vanadium. The most difficult aspect of remediation of this mining waste and acid rock drainage site was that the town of Uravan was in the middle of the site. The final remediation solution required moving everyone, and all buildings, from the town permanently (USEPA 2011b).

larger cities and towns. Smelting, which involves heating in the presence of flux, is employed to melt the mineral concentrates and segregate metals from their combined elements (e.g., oxygen or sulfur). Refining is a more advanced process to segregate one metal from another and produce a higher-grade, refined metal or metallic compound that can be sold for use in metal applications. Refining of the recovered metal can involve pyrometallurgical, chemical, and electrochemical processes that must be properly managed to prevent negative environmental impacts from wastes or emissions. Smelting and refining typically produces a waste slag (mostly melted minerals and other materials that form a rough, glass-like substance when cooled) that can require special handling. These processes also produce flue gases or smoke-stack emissions that need to be sampled and monitored. Sulfur dioxide, mercury vapor, and smelter dusts can be produced, which should be properly managed to prevent air- and drainage-quality impacts. Sulfuric acid is commonly produced and sold as a by-product of sulfide smelting and roaster operations.

2.4.4 Mine Expansion

Many mines undergo one or more expansions during their operating life as economic conditions change because of factors such as changes in metal prices, development of new technologies, and discovery of new ore reserves. The challenge is to keep the existing mine operating efficiently while the permitting process to expand mining moves forward. Ultimately, the ease in permitting expansions can depend on the ownership of land, past compliance record, attitudes of the general public, and the policies of regulating agencies.

2.4.5 Standby/Inactive Mines

Many mines will be forced to temporarily cease production for a variety of reasons, mostly related to economic conditions. Monitoring is highly recommended during temporary closure. At this time, few people are on-site to visually monitor potential problems, and the local community generally is not as focused on the mining operation. The possibility of oversight by regulatory agencies increases because this can be the mine's time of highest vulnerability. Mine sites can be inactive for many years, depending on government regulations and economic conditions. It is important that funding for continuation of sampling and monitoring is maintained during these standby periods. Another situation that can occur is when environmental targets become more restrictive under the new permit, and mine operators have to be ready with analytical results to show that all past waters and wastes will meet the new permit levels.

2.4.6 Sampling and Monitoring During the Operations Stage

The operational units of mining and metallurgical processing can include the mine, overburden and waste-rock storage facilities, development waste materials, ore stockpiles, tailings, crushing facilities, concentration facilities, leaching facilities, tailings storage, and other operational facilities. Each of these units can require environmental monitoring, depending on their potential contaminants, process requirements, and their handling and closure characteristics. Potential environmental effects of mineral production include the following:

- Impacts to wildlife and fisheries
- Habitat loss
- Changes in local water balance
- Sedimentation
- Introduction of potential contaminants in tailings storage facilities or leaching-solution impoundments
- Formation of potentially unstable dams and highwalls
- Construction of tailings ponds or leaching pads and ponds
- Potential acid generation from ore
- Waste rock
- Pit walls and tailings
- Metal leaching by acid, neutral, and alkaline drainage
- Cyanide-solution contamination at gold and silver metallurgical recovery operations
- Wind-borne dust
- Gas and dust emissions from metallurgical operations
- Introduction of contaminants from blasting operations

Most of these impacts are routinely addressed or resolved as part of the facilities design during development and production stages.

An important aspect of the operations stage is compliance monitoring. One key to compliance monitoring is to have complete chemical analyses with analytical detection limits and other QA/QC (quality assurance/quality control) documentation. Complete analytical data can be used to verify predictions made during exploration and mine development, and can provide a basis for modifying plans during the operations stage (Kuipers et al. 2005, 2006). For example, selective handling of ore and waste material can become necessary during the operations stage to manage pyrite-bearing material and prevent or minimize acid mine drainage. Modeling of groundwater flow can require chemical data that were obtained during the exploration or mine development stages. Similarly, chemical analyses of bench samples of blastholes during open pit production can be useful in characterizing ore stockpiles and mine waste facilities for operation control as well as for reclamation. It is good practice to set site-specific threshold levels that trigger additional sampling or monitoring activities to ensure that compliance levels are not exceeded. Although many mines are not required to monitor nitrogen compounds or organics in water, it is good practice to include ammonia and nitrate analysis to monitor the effects of blasting, and to include TCE (trichloroethylene) analyses to monitor residues from degreasing of equipment in shop areas.

Sampling and monitoring during operations is also done to meet objectives during closure. Closure plans should be continuously reviewed, and results of sampling and monitoring should be compared to what was predicted for closure and appropriate changes should be made to the closure plan.

It can be important to characterize all materials, including backfill materials and dewatering fluids, using chemical, physical, and mineralogical techniques, and to maintain detailed records of sampling and analyses. At the Pinto Valley mine in Arizona, a tailings impoundment failed when waste rock was placed on tailings (Kiefer 1998). Supplementary geotechnical characterization and monitoring of the physical properties of the tailings and impoundment could have prevented this failure.

The water balance determined during the mine development stage needs to be continuously updated during the operations stage as new data become available (O'Kane and Waters 2003; Marcoline et al. 2003). USEPA found that a common thread among past Comprehensive Environmental Response, Compensation, and Liability Act (commonly called Superfund) mine sites is that they did not have an accurate water balance (Russell 2006).

Expansions and temporary closures can occur during mine operation. Frequent reexamination of the sampling and monitoring programs and revision of sampling plans are ways to lower the impacts of these circumstances. It is good practice to maintain monitoring programs during temporary mine closures and to maintain adequate staff and funding to be able to address unforeseen problems.

2.5 CLOSURE/POSTCLOSURE

Today, numerous mines and metallurgical facilities in many countries have preliminary mine-closure plans in place prior to development or operations. It is recommended that closure plans begin at the start of mining, if not during the exploration phase (H. Jones 2011), and that plans are updated on a regular basis. *Closure* generally is the period following excavation and operations during which activities must be conducted to complete mineral extraction and return lands to their intended postmining use. For some portions of the exploration and operation, closure can be done immediately. For example, a waste rock pile or heap leach pad that is no longer being used can be closed. During exploration activities, drill pads and excavation pits can be closed. Wherever possible, closure activities should be planned concurrently with exploration and operations. This overcomes the cost of moving material more than once and is, thereby, cost-effective because operations handle some of the closure costs, which can also offer other economic benefits, such as taxation, in some jurisdictions. Closure plans usually require that mine and process structures be removed, land be graded and revegetated, openings be sealed or safeguarded, and lands be generally returned to a beneficial use other than mining. *Postclosure* is that period during which local environmental equilibration takes place (e.g., groundwater and surface-water conditions return to steady state and vegetation reestablishes). During both the closure and postclosure stages, monitoring can be required, and continuing monitoring after closure is proactive.

Closure can involve recontouring of pit walls and waste-rock storage facilities, covering of reactive tailings and waste-rock storage facilities, decommissioning of roads, dismantling of buildings, re-seeding/planting of disturbed areas, ongoing monitoring, and possible water quality treatment and discharge. Geomorphological stability in the long term is a prime objective of postclosure. For example, Table 2.1 summarizes the differences between natural hills and artificial rock piles and tailings (H. Jones 2011). Some of the potential problems include seepage of contaminated solutions into groundwater and surface water from MIW, wildlife and fisheries habitat loss, re-vegetation failure, wind-borne dust, and slope and tailings impoundment failure. Plugging of shafts and adits needs to be carefully evaluated because plugging can cause changes in the groundwater flow and possibly contribute to contamination of the groundwater. Closure

TABLE 2.1 Comparison between natural hills and waste rock piles/process tailings

Element	Mine Waste-Rock Piles and Process Tailings	Natural Hills
Construction material	Loose, uncompacted run-of-mine waste or run-of-mill waste	Bedrock core with weathered and or eroded regolith
Top surface	Varying size, large to small; surface typically bladed flat	Varying size, large to small; domed
Slope angle	Uniform, typically at angle of repose	Concave to convex, typically less than angle of repose
Slope configuration	Large rock piles and tailings constructed with regular steps; smaller rock piles, single slopes	Smooth
Drainage	Poorly drained	Well-developed, clearly defined drainages
Vegetation	Uniform age, introduced by the company	Mature and varied for climate and topography

Source: Adapted from H. Jones 2011.

plans typically establish a postmining use of the mine site, except for uranium mill tailings. Some sites may require long-term sampling and monitoring, whereas other sites will require no monitoring. Uranium mill tailings in the United States are controlled by specific federal regulations that provide for 1,000 years of monitoring and control by the federal government; these areas will not be released for future use of any kind. Biological media and water are the major media sampled during the final stages of the mine. Mine closure plan updating should be considered in all phases of mining.

2.5.1 Sampling and Monitoring During the Closure/Postclosure Stage

Closure plans typically establish a postmining use of the mine site commensurate with the desires of the landowner and adjacent local land uses. Some sites can require long-term maintenance and monitoring. Important factors in the closure/postclosure stage are to maintain monitoring and not to discontinue it too soon. For example, Placer Dome plans to monitor Las Cristinas mine in Venezuela for potential ARD for at least 20 years after mine closure because there are extensive underground and surface pathways for ARD, and any problems would have to be mitigated quickly (Dahlberg 1999). Problematic conditions can arise during, or years after, closure. At the San Luis mine in southern Colorado, a metal-rich seep developed adjacent to a backfilled pit during closure. Even though operations had ceased, there were personnel and equipment on-site to handle the problem, and mitigation was able to begin quickly. Continued monitoring can lend insight into source-control possibilities in lieu of ongoing treatment of MIW.

Continued characterization of solid materials moved during contouring and reclamation can avert unexpected results during closure. Also, predictive tests need to be reevaluated to assess their continued reliability. Some postmining uses, such as property sales or sale of non-acidic waste rock for construction purposes, could generate income for continued monitoring of the site.

2.6 HISTORICAL, INACTIVE, AND ABANDONED MINES

Historical, inactive, and abandoned mines, also known as legacy mines, offer the greatest challenge to the mining and environmental industries. These mines are historical, typically old, have not been mined for many years, and some of them do not have an owner and/or the original miner is unknown (i.e., true abandoned mines). Information on these sites, in many cases, is limited or nonexistent, and many would not comply with today's high mining standards of operation. Many of more than 500,000 legacy mine sites in the western United States pose no safety or environmental risk, but some of the mines could be challenging to remediate.

Historical miners had different operational philosophies than do miners of today. These miners had to make a profit quickly; often they would selectively mine only the high-grade ore (i.e., mine the high-grade ore without taking the lower-grade material). Exploration methods commonly involved construction of pits, shafts, and/or adits that have not been utilized in production of any ore. Historical miners typically located ore processing facilities (mills, tailings, stockpiles, etc.) near the production or haul shaft. Processing spoils were dumped near the processing unit, in many places into arroyos, streams, rivers, lakes, or other drainages. There were no laws or other guidelines available to prevent contamination of nearby drainages. Monitoring programs were nonexistent; sampling was conducted only to determine zones that could be economically mined. Metal mines were developed by underground and surface mining methods that were as inexpensive as possible, and safety and stability were not top priorities. Historical mines had as few ventilation shafts as possible in order to save money, resulting in poor air circulation. As little groundwater as possible was pumped from the mine. Mine surveying and mapping was minimal to nonexistent. Upon depletion of the ore, many mines were abandoned; mine waste piles and tailings piles were not engineered and were left as they were created without caps or vegetative cover. Surface and excess equipment was typically sold for scrap. It is important to understand the exploration, development, and production history of any mine before attempting to reclaim these sites (see case studies in Appendix 5).

Drainage quality is adversely affected in many past mining districts, but because there are typically no baseline data on conditions prior to mining, the cause and extent of the drainage quality is suspect. It is difficult to determine natural drainage conditions from those caused by mining. Studies determining background conditions in mining districts by USGS, such as the Questa and Red River districts in New Mexico (Nordstrom 2007) and Animas Basin in Colorado (Church et al. 2007b, 2007c), can be used as analog examples (see Appendix 5). The use of GEMs can provide some general information on basic deposit types (Appendix 6). What are considered poor mining practices today, such as dumping mine wastes into drainages, were considered a common, economic mining practice in the past, and these sites can still affect drainage quality.

One of the most difficult tasks in remediating historical mine sites is determining and characterizing the baseline conditions or natural background that existed prior to mining (Nordstrom 2007; Church et al. 2007a). Baseline conditions are necessary to establish meaningful remediation goals and monitoring programs. Methods used for this evaluation include the following:

- Integration of historical information
- GEMs
- Data from unmined, mineralized areas
- Analyses of water from monitoring wells (Figures 2.4 and 2.5)
- Leaching studies
- Statistical analyses
- Isotopic studies
- Identification of background by subtracting mining influences
- Computer modeling

Courtesy of Virginia T. McLemore.

FIGURE 2.4 Monitoring well near the Copper Flat pit lake (porphyry copper deposit), Hillsboro District, New Mexico

2.7 SUMMARY

The mine life cycle can be divided into four main stages: exploration, mine development, operations, and closure/postclosure (Figure 2.1). Information and data collected during exploration can be applied to later stages of the mine life cycle. More ground is sampled during the exploration stage than is mined and can produce a regional geochemical and environmental picture that can be applied to environmental planning and monitoring. Permitting and mining can proceed to the second stage of the mine life cycle, mine development, once an economic mineral deposit is discovered, its limits defined, and the prefeasibility/feasibility studies indicate that an economic deposit exists.

Mine development is the transition stage between exploration and mine operation. Hence, many issues considered during exploration also need to be addressed, sampled, and monitored during this stage of mining. Permitting, which generally takes place during the mine development stage, can require significant environmental data to accommodate the permitting process.

The operations stage of mining follows and includes the process whereby ores and waste rock are removed from the ground, milled and/or leached, and concentrated into a salable product, such as a metal sulfide concentrate or high-grade gold and silver bullion. The operational units of mining and metallurgical processing can include the mine, overburden and waste-rock storage facilities, development waste materials, ore stockpiles, tailings storage facilities, crushing facilities, concentration facilities, leaching facilities, and other operational facilities. Each of these units can require sampling and monitoring, depending on potential contaminants and their handling and closure characteristics.

The end of the mine life cycle is closure/postclosure, and closure plans typically establish a postmining use of the mine site commensurate with the desires of the landowner and adjacent local land uses. Some sites can require long-term maintenance and monitoring. Historical, inactive, and abandoned mines, also known as legacy mines, offer the greatest challenge to the mining and environmental industries for reclamation.

Courtesy of Virginia T. McLemore.

FIGURE 2.5 Monitoring field, Ambrosia Lake uranium subdistrict, Grants District, New Mexico

2.8 REFERENCES

Alaska Department of Environmental Conservation. 2009. Inspection report. http://dnr.alaska.gov/mlw/mining/largemine/kensington/pdf/dec/101509ir.pdf. Accessed 8/23/13.

Bove, D.J. 1994. *Preliminary Report on the Geology of Hydrothermally Altered Areas Within the Upper Alamosa River Basin, Colorado.* U.S. Geological Survey Open-File Report OF 94-224. http://pubs.usgs.gov/of/1994/0224/report.pdf. Accessed 8/23/13.

Castendyk, D.N., and Eary, L.E., eds. 2009. *Mine Pit Lakes: Characteristics, Predictive Modeling, and Sustainability*. Management Technologies for Metal Mining Influenced Water series, Vol. 3. Littleton, CO: SME.

Church, S.E., Fey, D.L., and Unruh, D.M. 2007a. Trace elements and lead isotopes in modern streambed and terrace sediment—Determination of current and premining geochemical baselines. In *Integrated Investigations of Environmental Effects of Historical Mining in the Animas River Watershed, San Juan County, Colorado.* Edited by S.E. Church, P. von Guerard, and S.E. Finger. U.S. Geological Survey Professional Paper 1651. http://pubs.usgs.gov/pp/1651/. Accessed 8/23/13.

Church, S.E., Kirschner, F.K., Choate, L.M., Lamothe, P.J., Budahn, J.R., and Brown, Z.A. 2007b. *Determination of Premining Geochemical Background, Blue Creek Drainage, Stevens County, Washington.* U.S. Geological Survey Scientific Investigations Report 2007-5262. http://pubs.usgs.gov/sir/2007/5262/. Accessed 8/23/13.

Church, S.E., von Guerard, P., and Finger, S.E., eds. 2007c. *Integrated Investigations of Environmental Effects of Historical Mining in the Animas River Watershed, San Juan County, Colorado.* U.S. Geological Survey Professional Paper 1651. http://pubs.usgs.gov/pp/1651/. Accessed 8/23/13.

CIM. 2005. CIM definition standards. web.cim.org/UserFiles/File/CIM_Definition_Standards_for_Mineral_Resources_and_Mineral_Reserves.pdf. Accessed 8/23/13.

Coeur Mining, Inc. n.d. Kensington, Alaska, 100% gold operation. www.coeur.com/operations/kensington-alaska#.UbYnp6BUHTo. Accessed 8/23/13.

Dahlberg, K. 1999. 14 steps to sustainability. Mining Report Card, Placer Dome, Inc. Washington, DC: Mineral Policy Center. www.earthworksaction.org/files/publications/14stepstosustainability.pdf. Accessed 8/23/13.

Diehl, S.F., and Schuster, R.L. 1996. Preliminary geologic map and alteration mineralogy of the main scarp of the Slumgullion landslide. In *The Slumgullion Earth Flow: A Large-Scale Natural Laboratory.* Edited by D.J. Varnes and W.Z. Savage. U.S. Geological Survey Bulletin 2130. http://pubs.usgs.gov/bul/b2130/. Accessed 8/23/13.

Dowd, P.A. 2002. The assessment and analysis of financial, technical, and environmental risk in mineral resource evaluation. In *Deposit and Geoenvironmental Models for Resource Exploitation and Environmental Security*. Edited by A.G. Fabbri, G. Gaal, and R.B. McCammon. Netherlands: Kluwer Publishers. pp. 187–221.

Golden, K. 2008. Acid drainage found at Kensington mine: Juneau Empire, September 30, 2008. www.juneauempire.com/stories/093008/loc_338558528.shtml. Accessed 8/23/13.

Gray, J.E., Coolbaugh, M.F., Plumlee, G.S., and Atkinson, W.W. 1994. Environmental geology of the Summitville mine, Colorado. *Econ. Geol.* 89:2006–2014.

Hartman, H.L. 1992. Introduction to mining. In *Mining Engineering Handbook*. Edited by H.L. Hartman. Littleton, CO: SME. pp. 3–42.

Jones, H. 2011. Mine closure: Plan for the end from the beginning. *Min. Eng.* 63(5):107–109.

Jones, W.R. 2007. History of mining and milling practices and production in San Juan County, Colorado, 1871–1991. In *Integrated Investigations of Environmental Effects of Historical Mining in the Animas River Watershed, San Juan County, Colorado*. Edited by S.E. Church, P. von Guerard, and S.E. Finger. U.S. Geological Survey Professional Paper 1651. http://pubs.usgs.gov/pp/1651/. Accessed 8/23/13.

JORC (Joint Ore Reserves Committee). 2012. *Australasian Code for Reporting of Exploration Results, Mineral Resources and Ore Reserves (JORC Code)*. www.jorc.org/docs/jorc_code2012%284%29.pdf. Accessed 8/23/13.

Kiefer, M. 1998. Cleaning up the Creek: A mining company mucks out its own mess at Pinto Creek. *Phoenix News Times*, May 7.

King, T.V.V., ed. 1995. *Environmental Considerations of Active and Abandoned Mine Lands, Lessons from Summitville, Colorado*. U.S. Geological Survey Bulletin 2220. http://pubs.usgs.gov/bul/b2220/b2220.pdf. Accessed 8/23/13.

Kuipers, J., Maest, A., MacHardy, K., and Lawson, G. 2005. Evaluation of the National Environmental Policy Act (NEPA) process for estimating water quality impacts at hardrock mine sites. SME Preprint 05-89. Littleton, CO: SME.

Kuipers, J.R., Maest, A.S., MacHardy, K.A., and Lawson, G. 2006. *Comparison of Predicted and Actual Water Quality at Hardrock Mines: The Reliability of Predictions in Environmental Impact Statements. Kuipers and Associates and Buka Environmental*. http://pebblescience.org/pdfs/ComparisonsReportFinal.pdf. Accessed 8/23/13.

Lord, D., Ethridge, M., Wilsom, M., Hall, G., and Uttley, P. 2001. *Measuring Exploration Success: An Alternative to the Discovery-Cost-per-Ounce Method of Quantifying Exploration Effectiveness*. Society of Economic Geologists, Newsletter No. 45, April.

Marcoline, J.R., Beckie, R.D., Smith, L., and Nicol, C.F. 2003. Mine waste rock hydrogeology: The effect of surface configuration on internal water flow. In *Proceedings, Sixth International Conference on Acid Rock Drainage*, Cairns, Queensland, Australia, July 14–17. Victoria, Australia, Australasian Institute of Mining and Metallurgy. pp. 911–918.

McCarthy, P.L. 2010. Objectives of feasibility studies. AMC Consultants internal paper. www.scribd.com/doc/91679597/Objectives-Feasibility-Studies. Accessed 8/23/13.

McLemore, V.T., ed. 2008. *Basics of Metal Mining Influenced Water*. Management Technologies for Metal Mining Influenced Water series, Vol. 1. Littleton, CO: SME. www.smenet.org/store/mining-books.cfm/Basics-of-Metal-Mining-Influenced-Water/259-8. Accessed 8/23/13.

McLemore, V.T. 2009. *Geologic Setting and Mining History of the Questa Mine, Taos County, New Mexico*. Open-File Report 515. Socorro, NM: New Mexico Bureau of Geology and Mineral Resources. http://geoinfo.nmt.edu/publications/openfile/details.cfml?Volume=515. Accessed 8/23/13.

Miller, R.A., and Hertel, T.M. 1997. Mine rock characterization: Zortman and Landusky mines, Little Rocky Mountains, Phillips County, North-Central Montana. In *Proceedings of the Fourth International Conference on Acid Rock Drainage*, Vancouver, BC, Canada, May 31–June 6. pp. 515–532.

NI 43-101. 2011. Rules and policies. In *Standards of Disclosure for Mineral Projects*. 34 OSCB 7086. www.ccpg.ca/profprac/en/Documents/ni_20110624_43-101_mineral-projects.pdf. Accessed 8/23/13.

Nordstrom, D.K. 2007. *Questa Baseline and Pre-Mining Ground-Water Quality Investigation. 25. Summary of Results and Baseline and Pre-Mining Ground-Water Geochemistry, Red River Valley, Taos County, New Mexico, 2001–2005*. U.S. Geological Survey Professional Paper 1728. http://pubs.usgs.gov/pp/1728/pdf/p1728_508.pdf. Accessed 8/23/13.

O'Kane, M., and Waters, P. 2003. Dry cover trails at Mt. Whaleback—A summary of overburden storage area cover system performance. In *Proceedings, Sixth International Conference on Acid Rock Drainage*, Cairns, Queensland, Australia, July 14–17. Victoria, Australia: Australasian Institute of Mining and Metallurgy. pp. 147–153.

Pendleton, J.A., Posey, H.H., and Long, M.B. 1995. Characterizing Summitville and its impacts: Summarizing the scene. In *Summitville Forum Proceedings*. Edited by H.H. Posey, J.A. Pendleton, and D. Van Zyl. Colorado Geological Survey Special Publication 38. pp. 1–12.

Plumlee, G.S., and Nash, J.T. 1995. Geoenvironmental models of mineral deposits—Fundamentals and applications. In *Preliminary Compilation of Descriptive Geoenvironmental Mineral Deposit Models*. U.S. Geological Survey Open-File Report 95-0831. http://pubs.usgs.gov/of/1995/ofr-95-0831/. Accessed 8/23/13.

Russell, C. 2006. Mine drainage challenges from A to Z in the United States. In *Proceedings, 7th International Conference on Acid Rock Drainage (ICARD 7)*. Lexington, KY: American Society of Mining and Reclamation. pp. 1785–1802. www.asmr.us/Publications/Conference%20Proceedings/2006/1785-Russell-CO.pdf. Accessed 8/23/13.

Schmiermund, R.L., and Ranville, J. 2006. Application of geo-environmental and biotic ligand models in the life cycle of mine development—A flowchart for integration. In *Proceedings, 7th International Conference on Acid Rock Drainage (ICARD 7)*. Lexington, KY: American Society of Mining and Reclamation. pp. 1920–1925. www.imwa.info/docs/imwa_2006/1920-Schmiermund-CO-3.pdf. Accessed 8/23/13.

Schmiermund, R.L., Lazo, M.C., and Parnow, C.C. 2006. Application of geo-environmental models to accelerated EIA and permitting processes for an Andean porphyry Cu-Au deposit. In *Proceedings, 7th International Conference on Acid Rock Drainage (ICARD 7)*. Lexington, KY: American Society of Mining and Reclamation. pp. 1903–1911. www.imwa.info/docs/imwa_2006/1903-Schmiermund-CO-1.pdf. Accessed 8/23/13.

Semple, R.B., Jr. 1995. Canceling the New World mine. *The New York Times*, 10 December.

USEPA (U.S. Environmental Protection Agency). 2011a. Summitville mine, Del Norte, Colorado, Superfund Program (EPA ID: COD983778432). www2.epa.gov/region8/summitville-mine. Accessed 8/23/13.

USEPA (U.S. Environmental Protection Agency). 2011b. Uravan Uranium Project (Union Carbide), Uravan, Colorado, Superfund Program (EPA ID: COD007063274). www2.epa.gov/region8/uravan-uranium-project-union-carbide. Accessed 8/23/13.

Vincent, K.R. 2008. *Questa Baseline and Pre-Mining Ground-Water Quality Investigation. 17. Geomorphology of the Red River Valley, Taos County, New Mexico, and Influence on Ground-Water Flow in the Shallow Alluvial Aquifer*. U.S. Geological Survey Scientific Investigations Report 2006–5156. http://pubs.usgs.gov/sir/2006/5156/. Accessed 8/23/13.

CHAPTER 3

Sampling Considerations in the Mining Environment

Kathleen S. Smith, Virginia T. McLemore, and Carol C. Russell

(with contributions by Andrew S. Todd, James J. Gusek, Lisa Bithell Kirk, Karl L. Ford, and Susan J. Caplan)

3.1 INTRODUCTION

The intent of this chapter is to highlight some information and issues that relate to sampling at mining and processing sites, and to discuss some relevant information from recent research studies. Internet sources related to this chapter are listed in Section 3.13. Appendices 2 and 3, which summarize many types of sample-collection and analytical methods, are supplementary and complementary to the information herein. In addition, Appendix 1 lists sources for online resources related to sampling, and Appendix 5 consists of several case studies that illustrate different aspects of sampling and monitoring. Sampling and monitoring of pit lakes are discussed in volume 3 of the Management Technologies for Metal Mining Influenced Water series, *Mine Pit Lakes: Characteristics, Predictive Modeling, and Sustainability* (Castendyk and Eary 2009); considerations related to mitigation of mining influenced water (MIW) are covered in volume 2, *Mitigation of Metal Mining Influenced Water* (Gusek and Figueroa 2009); prediction methods are included in volume 5, *Techniques for Predicting Metal Mining Influenced Water* (Williams and Diehl 2014); and geochemical modeling is covered in volume 4, *Geochemical Modeling for Mine Site Characterization and Remediation* (Nordstrom et al., forthcoming). The general media types of water, solids, biological (aquatic), and air are included in this chapter. Table 3.1 lists many of the types of data and sampling media that can be considered for inclusion in sampling and monitoring programs.

Metals present unique sampling issues in that they are naturally occurring and have variable content in solids and variable concentrations in MIW from different geographic areas. In addition, some metals are essential nutrients but can be toxic at higher concentrations in a variety of media. Also, the form of a metal can control its mobility, bioavailability, and toxicity, and metal-ion mixtures can have additive or synergistic toxicological effects. Because metals generally occur in mixtures at mining and processing sites, there can be several pathways and receptors of interest due to the potentially dissimilar behaviors of different metals. These factors make sampling for metals at mining sites a challenge and underscore the importance of having a fundamental understanding of the physicochemical and biogeochemical processes that control metal behavior. Nordstrom (2011) provides an overview of hydrogeochemical processes that control metal mobility in mined and mineralized systems.

At the outset of a sampling program, it is essential to define the reasons for sampling, the questions to be answered, the desired analytes and sampling media, and the desired degree of

TABLE 3.1 Data and sampling media for sampling and monitoring programs

Physical	
Topography and altitude	Stream/river flow data
Natural drainage vs. human-made drainage	Slope aspects
Geological	
Structural geology (faults, fractures, joints, lineaments)	Geomorphology
Seismic data/earthquake prediction	Deposit type
Stratigraphy and lithology	Mining methods
Particle size analysis	Mineralogy
Aquifer Characteristics	
Permeability/porosity	Groundwater flow direction and rates
Hydraulic conductivity	
Climatological	
Temperature	Growing season
Evapotranspiration	Days of sunlight
Rainfall/snowfall	Floods (10 yr, 100 yr, etc.)
Freeze-thaw	Storm events and types
Wind direction and speed	Probable maximum precipitation event
Solid	
Overburden	Tailings
Country rock	Soils/topsoil
Ore	Subsurface samples (drill core, cuttings, other)
Sub-ore or low-grade material stockpiles	Sediments (lake, stream, river, overbank)
Sludge	Spent leach ore/material
Waste (piles, other)	
Liquid	
Surface water Storm water (impacted and non-impacted) Stream/river Lakes and ponds Processed wastewater Other wastewater	Human-made discharge points (adits, tunnels, pit walls, etc.) Rainwater/snow melt Brine Residuals from passive treatment Process solutions
Groundwater	Pit lakes
Leachate	Seeps
Surface runoff	Springs
Air	
Gases (oxygen flux, cyanide, mercury)	Aerosols
Particulates (dust)	Fugitive dust
Biological	
Humans	Habitats
Animals Aquatic (macroinvertebrates, vertebrates, fish) Terrestrial	Plants Aquatic (algae) Terrestrial
Microorganisms (bacteria) Aquatic Terrestrial	

confidence in the answer(s). Numerous textbooks, compilations, handbooks, and articles address the types, quantity, and quality of data (e.g., Keith 1991; Horowitz et al. 1994; USEPA 1984, 2000, 2001a, 2001d, 2003b, 2006c, 2007b, 2007c, variously dated; Tremblay and Hogan 2001a, 2001b; Zhang 2007; Rice et al. 2012; ICMM 2008; INAP 2009; GRI 2010; USGS variously dated; BCMOE 2012). The U.S. Environmental Protection Agency (USEPA) provides links (listed in Section 3.13) to several methods for environmental sampling. Callender (2003) reviews sources and distribution of heavy metals in a variety of media.

USEPA has also developed an agency-wide Quality System that includes a series of guidance documents (see Section 3.13; USEPA 2002a). USEPA's data quality objectives (DQOs) guidance document (USEPA 2006b) describes the DQO process, which is USEPA's recommended planning procedure when environmental data are used to select between two alternatives or to derive an estimate of contamination. USEPA's Quality System is discussed in more detail in Chapter 4 of this handbook.

There has been an evolution in the approach for characterization of hazardous waste sites, which may also be applied to environmental sampling and monitoring of mining and processing sites. Traditional sampling and analysis programs have been relatively inflexible, and data quality has been focused on analytical error, with little concern for other factors (e.g., sampling techniques) that can influence data quality. To address some of these issues, Crumbling et al. (2001) introduced the Triad approach, which is an integrated work strategy to manage decision uncertainty that allows consideration of all parameters that influence data quality. Triad is a federal and state interagency partnership (see Section 3.13). The Triad approach relies on thorough, systematic planning to set clear project goals. The three elements of the Triad approach are (1) systematic (or strategic) project planning, (2) dynamic work strategies, and (3) real-time measurement technologies. This approach emphasizes the need for a multidisciplinary team of experts for planning and decision making. Dynamic work strategies allow work planning documents to be written in a flexible way so that the course and decisions of the project can adapt in real time as new information becomes available. The Triad approach allows for a variety of analytical techniques. If possible, initial analytical measurements should be made in the field (real-time) so that they can inform the dynamic work strategies while the crew is still in the field. Some techniques that can support real-time measurements include field kits, field analytical instruments, in situ sensing systems, field leaching tests, and some geophysical techniques. The Triad approach reinforces the distinction between data quality and analytical quality, and considers sampling quality to be just as important to data quality as is analytical quality. Data quality can be achieved by initially using results from more rapid, often less expensive, analytical methods to provide high-density data that can assess sample representativeness. The level of tolerable error depends on the objectives of the investigation. Although the Triad approach has yet to be adopted for mining sites, screening level approaches have been utilized to assess potential environmental considerations at mining sites. For example, Smith et al. (2002) discuss tools for rapid screening and characterization of waste piles, and Jordan (2009) outlines an approach by which geological information can be used in the environmental assessment of mining areas. Schmidt et al. (2012) and Schmidt (Appendix 5, this handbook) discuss the links between geology and potential impacts on aquatic ecosystems. Once sample representativeness is assessed, more precise and accurate analytical techniques can be used if necessary.

Geo-environmental mineral deposit models (GEMs; also discussed in Chapter 2 and in Appendix 6) are a geology-based classification system for identifying environmental challenges associated with specific types of mineral deposits and can provide an initial prediction of water quality at a given mine site based on the type of mineral deposit (e.g., Plumlee et al. 1994, 1995,

1999; Appendix 6). In addition to being useful in developing a site conceptual model, GEMs are helpful as a guide in forecasting and prioritizing potential environmental considerations and challenges at a particular site, in assessing potential remediation options, and as a reality check for evaluating numerical modeling results. Information about GEMs is provided by du Bray (1995), Plumlee and Nash (1995), Plumlee (1999), Plumlee et al. (1999), Wanty et al. (1999), Seal and Foley (2002), Kwong (2003), Seal and Hammarstrom (2003), Schmiermund et al. (2006), Schmiermund and Ranville (2006), and Seal et al. (2009).

3.1.1 Background, Baseline, and Premining Conditions

A topic of potential confusion that deserves mention is the difference between the terms *background* and *baseline*. These words are often used interchangeably, but many researchers make a distinction between them. Therefore, when discussing background or baseline conditions, it is critical to define the terms in the context that they are being applied. Lee and Helsel (2005) define the term *environmental background* as being descriptive of the natural tendency of an environmental system, or material, in the absence of human influences. In contrast, they define an *environmental baseline* as a summary of existing conditions over some time frame for some environmental system or material of interest. They also state that baselines can, and typically do, include influences of human activities such as land use or mining and processing. Galuszka (2007) defines *baseline* as a measure of a given sample in a specific location and time. The term *natural background* has been widely used to refer to background levels that are not influenced by human activities. Runnells et al. (1992) suggests that *natural background* conditions be the remediation goal at environmentally impacted mining sites. USEPA policies require cleanup to *natural background conditions* (see USEPA 1997), where natural background is defined as concentration due only to nonanthropogenic sources. Reimann and Garrett (2005) introduced the term *ambient background* to describe "the unmeasurably perturbed and no longer pristine natural background." The European Chemicals Agency (ECA 2008) provides guidance on determining the background and baseline for metals. Because background usually represents conditions prior to any human influence, background conditions often cannot be accurately determined and are commonly determined based on conditions at analog sites that have not been subjected to human-induced changes, or calculated based on modeling or projections to past conditions. These types of projections involve establishing the spatial and temporal distribution of elements of interest, and determining their behavior in response to different physicochemical and biological conditions.

Premining conditions are those that existed prior to mining, regardless of other human influences. Estimation of premining conditions has been a topic of recent interest and research. For example, considerable work is being done to determine premining conditions at the Pebble mine in Alaska (Alaska DNR 2013). Mine sites are generally located in areas with naturally elevated metal concentrations in water, and metal contents of sediment, soil, vegetation, and other biota. Therefore, it is important to conduct careful sampling and monitoring before mining to document premining conditions at proposed mine sites (Alpers and Nordstrom 2000; McLemore et al. 2007). It also is important to develop scientifically defensible approaches and methods to estimate premining conditions at existing mine sites where no premining conditions were documented (e.g., Church et al. 2007a, 2007b). The U.S. Geological Survey (USGS), in cooperation with the New Mexico Environment Department and supported by Chevron Minerals Co. (formerly Molycorp, Inc.), conducted the *Questa Baseline and Pre-Mining Ground-Water Quality Investigation* (Nordstrom 2008) to estimate premining groundwater quality at the Questa molybdenum mine in northern New Mexico. USGS identified several geochemical trends and found

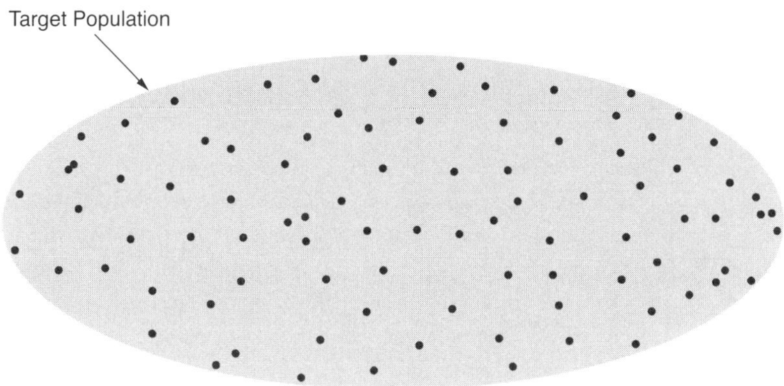

FIGURE 3.1 Illustration showing the relationship between a target population (gray oval) and a Multi Increment sample, represented by the black dots

that these trends, combined with mineral-solubility controls and site-specific lithology and mineralogy, provide constraints on premining groundwater quality (summarized in Nordstrom 2008; see USGS 2011). USGS conducted detailed studies in the Upper Animas River Basin, Colorado, which is an area with a legacy of MIW from historical metal production (see USGS 2013b). Characterization, monitoring, and process-oriented studies provide insights for effective remediation strategies and estimation of premining conditions (summarized in Church et al. 2007c; see case studies by Church and Williams in Appendix 5). Verplanck et al. (2009) provide an overview of controls on water quality in several areas draining different porphyry mineralized areas, and Runkel et al. (2007) present a simulation-based approach for estimating premining water quality.

3.2 SAMPLING OVERVIEW

Sampling is the process by which inference is made to the whole by examining a part. The sampling process should be defined by the DQOs (USEPA 2006b) and the sampling and monitoring program (see Figure 4.1 and discussion of international guidelines in Section 4.3, both in Chapter 4), which incorporates the project sampling plan, quality assurance plans (USEPA 2001b), and standard operating procedures (USEPA 2007c). The target population is that which is being sampled and is defined based on the objectives of the investigation. It is critical that the target population be clearly identified in detail at the beginning of the investigation, be understandable to the users, and that all of a given target population be equally available for sampling. Examples of a target population might include historical fluvial mill tailings within the 100-year flood plain of a stream, or fish along a particular stream reach, or groundwater contamination downgradient from an environmentally impacted mine site, or the top inch of soil in a particular area. It is likely that there will be multiple target populations for a given investigation.

A *sample* is that portion of the target population that is actually studied and used to characterize the target population. For the purposes of this discussion, a sample consists of several increments (subsamples) collected from the target population (Figure 3.1). Multi Increment (MI) sampling (EnviroStat, n.d.) is a sampling methodology that can be used to collect a representative sample and is discussed in more detail in Section 3.4.8.1.

The sample needs to be representative of the target population. The term *representative* commonly is loosely defined in qualitative terms. Determining *sample representativeness* involves careful planning and formulation of a proper sampling design. The representativeness of samples can be addressed by using either a probabilistic approach (e.g., Rollinson 1993; Koch and Link

WHAT DOES *REPRESENTATIVE SAMPLE* MEAN?

Kathleen S. Smith, U.S. Geological Survey, and Charles A. Ramsey, EnviroStat, Inc.

The term *representative sample* has been used for many applications. It is actually an operationally defined term that depends on the objectives of the investigation and the definition of the target population. A representative sample is both unbiased and precise within limits specified in the objectives. For a given investigation, the meaning of representative sample must be defined so that the ultimate decision-making needs of the investigation can be met. The concept of a representative sample exists throughout the life cycle of an investigation, from field sampling, to maintaining sample integrity, to sample preparation through staged comminution, to laboratory subsampling, to sample analysis. Determining sample representativeness involves careful planning and formulating a proper sampling design.

For field sampling, sample preparation, and laboratory subsampling of particulate materials, it all comes down to the sampling error. "There is always some degree of sampling error unless the entire population is analyzed" (Ramsey and Hewitt 2005). Pierre Gy (1982, 1998) studied the technical and statistical aspects of sampling error applied to mineral processing, and Pitard (1993) further refined Gy's sampling theory. Sampling error is related to the heterogeneity (both compositional and distributional) of the target population. Two primary sources of precision error are (1) the grouping and segregation error (GSE), which is related to distributional heterogeneity and number of subsamples collected to compose a sample; and (2) the fundamental sampling error (FSE), which is related to the compositional heterogeneity, particle size, and the mass of the sample. The major source of bias error can be minimized by using correct sampling practices and devices (see Pitard 1993). As a rule of thumb, Pitard (1993) recommends collecting at least 30 increments (subsamples) to constitute a sample in order to reduce the GSE relative to the FSE (in practice, 50–100 increments are usually required). Hence, when correct sampling practices are employed and sufficient increments are collected, then the majority of the total sampling error is due to the FSE.

The FSE can be estimated prior to sampling and can be reduced by increasing the sample mass or decreasing the particle diameter. The FSE can be estimated using the following equation (when the sample size is much smaller than the target population being sampled):

$$\sigma^2_{FSE} = (K\,d^3)/M_s$$

where M_s is the sample mass (g), d is the maximum particle diameter (cm), and K is an empirical constant for a given material that is related to the particle shape, range of particle sizes, particle mineralogical composition, and particle liberation.

To obtain representative samples, it is essential to be aware of the properties of heterogeneous materials and the guidance provided by sampling theory. Investigators need to proactively design sampling and analysis programs with clearly defined data quality objectives to achieve data of known quality and integrity (Wait and Ramsey 2007).

2002; Schreuder et al. 2004; USEPA 2006a) or a sampling theory approach (see Representative Sample sidebar and Sections 3.4.8 and 3.4.9). If using a probabilistic approach, a software tool called Visual Sample Plan (VSP) developed by the Pacific Northwest National Laboratory (Richland, Washington) can help develop a sampling plan that supports the DQO process (see Section 3.13 and Nuffer et al. 2009). VSP is founded on statistically based sampling designs. Decision Error Feasibility Trials (DEFT; USEPA 2001a) is another computer model that supports the DQO process (see Section 3.13). DEFT assists with the seven-step DQO process (USEPA

2006b) and is used to develop statistical sampling design plans that allow users to evaluate the financial feasibility of incorporating selected DQO constraints into a statistical sampling design before developing a final plan.

A sampling theory approach considers the effect of both sampling and analytical errors on the representativeness of samples (Crumbling et al. 2001; Gy 1982, 1998; Maney and Wait 2005; Pitard 1993; Ramsey and Hewitt 2005; Wait and Ramsey 2007). The major source of uncertainty in environmental data sets is due to the heterogeneity of environmental materials. Use of sampling theory can help minimize sampling errors that are due to the heterogeneity of the target population (see Sections 3.4.8 and 3.4.9).

Sampling methods are the procedures of selecting those sampling units that provide estimates of the whole with associated margins of uncertainty. Poor sampling can introduce sample bias as well as lack of precision. Uncertainty arises from examination of only a part and not the whole. For example, selecting a specific sample size fraction in characterizing sulfidic waste rock piles becomes important and a bias can be introduced because the acid-generating and acid-neutralizing minerals are not evenly distributed in the various size fractions. Similarly, surface sampling of environmentally impacted pit lakes can provide an incorrect estimate of contaminant loads if the pit lake is stratified with respect to dissolved constituents. In such cases, replicate sampling or increasing sample volume or mass will not correct the bias.

In summary, it is essential to carefully define the target populations during the sampling planning phase. There are four main groups of sampling media (solid, liquid, biological, and air) that are generally considered by a sampling and monitoring program. The types of media sampled will be determined by the anticipated sources, pathways, and receptors potentially impacted by mining and processing activities, as well as by other factors such as climate and regulatory requirements. It is important to consider and address the types of sample variation (e.g., spatial and temporal). It is helpful to consider use of indicator species, either aqueous or biological, to provide an early warning system for potential environmental impacts.

3.3 WATER SAMPLING CONSIDERATIONS

The purpose of collecting water and process-solution samples at mine and processing sites can be to assess water quality, provide data for contaminant loading calculations, use in toxicological testing, locate contaminant source(s), provide input for modeling efforts, establish baseline conditions, manage water treatment processes, or meet regulatory requirements. A general summary of water quality characteristics is discussed by Hem (1985), a compilation of the biogeochemistry of major world rivers is given by Degens et al. (1991), and acid rock drainage (ARD) sampling is described by Downing (2011). Some Web sites that discuss general characteristics of water are listed in Section 3.13. The National Water Quality Monitoring Council also maintains a Web site and offers training and information relevant to the sampling and monitoring of water.

MIW can have many possible characteristics (Nordstrom and Alpers 1999; see Appendix 3 for discussion of water constituents and field measurements). Schmiermund and Drozd (1997) list five common possible characteristics of MIW, which include low pH, high sulfate (SO_4^{2-}), high iron and aluminum, high non-iron metals, and high turbidity. Importantly, not all MIWs need to have all of these characteristics (e.g., not all MIW is acidic), and each characteristic often requires different sampling and monitoring techniques. When evaluating an existing MIW, care should be taken to fully characterize the system, both absolutely and in terms of an appropriate baseline or background (Schmiermund and Drozd 1997; Smith 2011). For example, MIW from a cyanidation processing operation discharge may be much different than from a MIW source that contains high-pH values and metal species present as anions instead of cations. Focusing characterization activities solely on regulated constituents often results in incomplete or incorrect characterization, which could lead to costly oversights and insufficient data for remediation planning.

Major advances have been made in developing *clean* sampling procedures and extremely sensitive analytical techniques for trace elements. However, sampling procedures that address and capture the natural variability of constituents in water need some further development; sampling procedures can frequently be the major source of error in the overall process. It is critical to use current, relevant, documented sampling protocols and to record how, when, and where samples were collected, preserved, and analyzed. Textbook preservation procedures and holding times may not always be applicable to MIW. Numerous protocols describe sampling, preservation, and analytical procedures (e.g., USGS, variously dated). Great care must be taken to ensure that water quality sampling procedures, sampling devices, and preservatives do not contaminate the sample, react with the sample, or interfere with the analytical method, and all such information needs to be recorded in the sampling plan documents. In addition, documentation of calibration should be included for all field instruments. When stream sediment is also being sampled, the water sample needs to be collected first to avoid contamination from the disturbed sediment. Experienced personnel should be chosen to collect water quality samples. Collection of quality assurance/quality control (QA/QC) samples, such as replicates and blanks, is very useful to determine if any contamination has occurred and whether variation in constituents between samples is due to contamination or to environmental variation. It is important to document such things as frequency of sampling; whether samples were composited; where in the stream samples were collected; whether samples were collected using a pump, a wire sampler, a composite sampler, or by a grab method; the existence of QA/QC samples; and whether samples were filtered (and filter size and type). It also is recommended that ambient weather conditions before and during the sampling event be recorded. For example, does the sampling event occur during a prolonged drought at the site or perhaps hours/days after the site received significant rainfall (see Section 3.3.6.1)?

Wire samplers are often used in metallurgical plants to collect shift composite samples. In this sampling procedure, a wire is passed through a constant stream of solution and a portion of the sample is diverted to flow down the wire to drip into a sampling bottle.

3.3.1 Filtration

Filtration and filtration techniques can affect the concentration of constituents of interest in the sample, so filtration should be specifically addressed in the sampling plan and procedures, and be consistent with the overall objectives of the investigation. Filtration through a 0.45-µm pore-size membrane is a widely accepted, regulation-driven method to separate particulates from water to obtain the *dissolved* fraction. However, this pore size is arbitrary and does not always remove all of the particulates from water. Examples of some particulates that may not be filtered include colloids (suspended solids, sometimes referred to as nanoparticles) and some microorganisms. Ranville and Schmiermund (1999) have presented a discussion and overview of colloids. Notably, the use of acid or other additives for preservation and the digestion of filtrates using acid or other chemicals likely dissolves at least some of the particulate material that passes through the filter.

Sampling of slurries, such as tailings, in metallurgical plants may require the use of a primary cross-stream sampler followed by one or more rotary Vezin samplers. The slurries are typically pressure filtered to separate the solids from the solution. Fine filtration can follow this preliminary separation, and gravity filtration in a funnel of small volumes of slurries is also a common practice for extraction of solution for metallurgical process control measurements.

Filters and filtration devices can be sources of contamination and need to be carefully precleaned when trace constituents are of interest. Also, there can be artifacts as a result of filtration (e.g., Danielsson 1982; Horowitz et al. 1992, 1996). For example, Horowitz et al. (1996) have illustrated that filter clogging changed the effective pore size of filters during filtration. When possible, it is good practice to filter samples in the field if filtration is necessary. Filtration of groundwater samples is discussed in Section 3.3.7.

3.3.2 Unstable Constituents and Field Measurements

Reduction-oxidation (redox) potential determination should not be considered to be a routine determination due to the difficulties inherent in its theoretical concept and its practical measurement (Nordstrom and Wilde 2005). Redox and Eh can be measured by several methods (see Appendix 3 for more detailed information). Eh determination using a platinum (or other noble metal) electrode is valid only when redox species are electroactive and present in the solution at concentrations greater than about 10^{-5} molal (Nordstrom and Wilde 2005). Generally, a single Eh measurement does not represent the entire system; in many MIWs, iron controls the response of the electrode because of its elevated concentration. Eh measurements can also be used to determine sulfide in sulfate-reducing systems (Nordstrom and Wilde 2005). Most redox species can be directly measured, which can be preferable to Eh measurements (see Baedecker and Cozzarelli 1992).

A number of parameters are traditionally measured in the field or processing plant rather than in the laboratory. These include temperature, specific conductance, pH, dissolved oxygen (DO), turbidity, and flow rate. Also, acidity/alkalinity samples need to be kept cool and be analyzed as soon as possible. Some measurements, such as turbidity, are designed to be measured in the field. When it is uncertain whether a particular constituent or parameter is stable over time, it is better to try to measure that constituent in the field. For example, in MIW resulting from ARD, the pH value should always be measured in the field because pH can change with time due to precipitation of hydrous iron oxides, evolution of carbon dioxide, and other issues. Field techniques are discussed in more detail in Appendix 3, and a review of field methods is given in Ficklin and Mosier (1999). Additionally, many types of speciation measurements must be performed in the field because the species being measured are not stable. When considering speciation sampling, it is cost-effective to first analyze for total concentration of the constituent to determine if the constituent is present at a concentration level that is sufficient for speciation analysis. Various field test kits are available for a number of constituents. These test kits can be very useful in determining unstable constituents or aqueous species and are discussed in more detail in Appendix 3.

3.3.3 Sensors and Passive Sampling Devices

Recent advances in sensor technology allow for greater capabilities in water quality monitoring. The ability to collect continuous, rather than discrete, data provides the foundation for many potential applications in sampling and monitoring that allow for better understanding of constituent variability. For example, continuous and remote monitoring are potential tools for determining total maximum daily loads (TMDLs), which are generally developed and monitored using discrete water samples collected on a fixed schedule; Walton-Day et al. (2012) discuss some of the hydrogeochemical complexities involved in developing TMDLs. Roig (2007) presents performance evaluation and validation information for several alternative screening methods (i.e., physicochemical devices, biosensors, bioassays), Glasgow et al. (2004) review advancements in sensor technology, Chapin and Todd (2012) describe an in situ sampling device that can improve sampling resolution, Wanekaya et al. (2008) provide an overview of several types of biosensors, and Horsburgh et al. (2010) discuss the use of surrogates to monitor constituents using sensor technology.

Passive sampling devices (e.g., diffusion gradients in thin films, stabilized liquid membrane devices) offer another alternative to discrete sampling. In passive sampling, the analyte concentration is integrated over the sampling period; passive samplers do not require an energy source and can be deployed in many types of environments (e.g., Brumbaugh et al. 2002; Balistrieri et al. 2007). However, passive sampling does not provide information about constituent variability. Vrana et al. (2005) review passive sampling techniques for environmental applications. Madrid and Zayas (2007) point out that one of the most important shortcomings of the use of sensors and passive sampling devices is validation under field conditions. See Appendix 3 for further discussion of various sampling devices.

3.3.4 Water Balance

It is essential to understand the water balance (also called water budget) in a given study area (see Argo Tunnel sidebar). Russell (2006) points out that USEPA has found a common thread among recently active Superfund mine sites: they did not have an accurate water balance. A water balance is the rate of change in water stored in an area (e.g., a watershed) balanced by the rate at which water flows into and out of the area. Water balances are fundamental to the conceptualization of hydrologic systems at all scales (Healy and Scanlon 2010.) Developing a water balance includes collection of hydrological (e.g., flow or discharge measurements, groundwater flow parameters) and climatological (e.g., precipitation, temperature, evapotranspiration) information. Some factors that can play an important role in a site-wide water balance include seasonal flow variability, runoff variability, evapotranspiration, lag discharge (e.g., storage in waste rock piles), mine dewatering, water application for dust control, process discharge retention in tailings facilities, and consumption. Healy et al. (2007) present an overview of water balances and provided several case studies. Water sources and losses need to be identified and, at least partially, quantified.

3.3.5 Groundwater/Surface-Water Interactions

Understanding interactions between groundwater and surface water is an important, often overlooked consideration in water management. Winter (1999) discusses the influence of surface-water bodies (e.g., streams, lakes, wetlands) on groundwater flow systems. Winter et al. (1998) and Idowu (2007) provide an overview of groundwater/surface-water interactions. They point out that groundwater flow systems can be local, intermediate, and regional in scale. Intermediate and regional flow systems have longer flow paths and contact time with subsurface materials, and these waters tend to contain dissolved constituents that reflect the composition of the subsurface materials (e.g., Wanty et al. 2009a, 2009b). Local flow systems can be very dynamic and have interactions with surface-water systems (see Figure 3.2; Toth 1963; Winter et al. 1998). Interactions at the surface-water/groundwater interface (the *hyporheic zone*) can play an important role in the concentration and load of constituents in surface water and can have significant environmental influences on biogeochemical processes (Harvey and Fuller 1998; Bencala 2000, 2005; Fuller and Harvey 2000). Gandy et al. (2007) provide a review of information about attenuation of potential contaminants in the hyporheic zone. Bencala et al. (2011) point out that streams gain and lose water over a range of spatial and temporal scales, and that it is important to consider the influence of scale on contaminant and nutrient fluxes. It is important to keep these surface-water/groundwater interactions in mind when designing a sampling plan and locating sampling sites (e.g., Harvey and Bencala 1993). The hyporheic zone is discussed in more detail in Appendix 3.

QUALITY AND CERTAINTY OF DATA—ARGO TUNNEL

Carol C. Russell, U.S. Environmental Protection Agency

An example of the need for an accurate water balance for an acid rock drainage (ARD) evaluation is clear when considering the Argo Tunnel, a 6.69-km (4.2-mile) tunnel completed in 1910 that passes from Idaho Springs (the portal) to underneath Central City in Colorado. The Argo Tunnel provided dewatering and ore haulage-ways draining more than 300 mines and is the largest single source of ARD and metals contamination to Clear Creek. As part of a Superfund remediation project, construction of a 700-gpm conventional sodium hydroxide treatment facility was completed in 1998. Although the plant has been successful in meeting the effluent quality goals, it has been at much higher costs than estimated due, in part, to inaccurate flow measurements of the ARD. Initial flows were measured at the portal, but the flows being captured several meters within the tunnel area are twice that of the initially measured flow (USEPA 2007d). Follow-up analysis of the flows found that fracture flow from the weathered surface near the tunnel exit and seasonal variations in flow are responsible for the differences in flow measurements. The plant was designed as a dual train system with one filter press. Rather than being able to run one train while the second was being maintained, both trains must function all of the time, necessitating more oversight and much more costly maintenance. Moreover, an additional filter press purchase incurred rush charges, and more filtration sludge than expected has led to increased disposal costs. Nevertheless, the plant continues to remove more than 453 kg (1,000 lb) of metal each day from the ARD before discharging to Clear Creek (CDPHE 2011). A document published by USEPA titled *Performing Quality Flow Measurements at Mine Sites* (USEPA 2001c) can assist in more accurate measurement of ARD from mines.

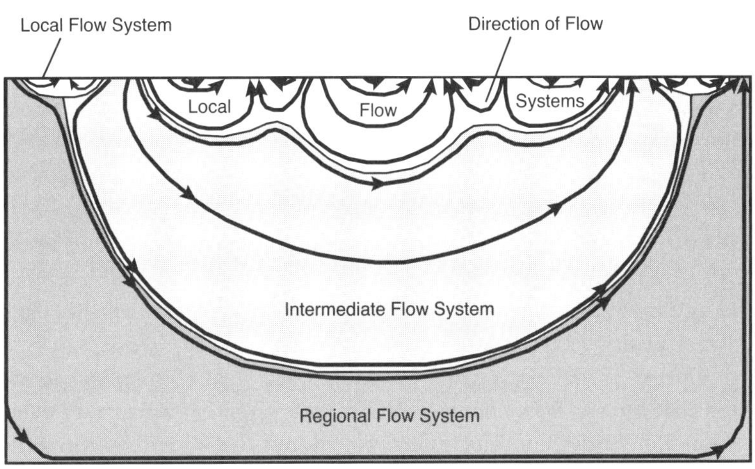

Source: Adapted from Toth 1963, as cited in Winter et al. 1998.

FIGURE 3.2 Different scales of groundwater flow systems: local, intermediate, and regional

3.3.6 Surface-Water Samples

Dissolved metals occur in many chemical forms, including free ions and inorganic and organic complexes. The chemical speciation of metals influences their bioavailability to and toxicological effects on aquatic organisms. Constituents in surface water draining mined and mineralized areas tend to be geochemically, spatially, and temporally variable, which presents challenges in collecting water quality samples. Important considerations in spatial variability include location of water body in the catchment, the underlying lithology, presence of weathering ore deposits or waste, process discharges, climate, and geochemical processes that can transform and redistribute the analytes. Important temporal considerations include seasonal, episodic, and daily changes. This section addresses some of the important considerations in collecting surface-water samples from mined and mineralized areas.

3.3.6.1 Temporal Considerations

Temporal variation of metal concentrations can be significant in streams draining mined as well as mineralized, unmined areas, and for streams impacted by excess and treated process water from mine dewatering, MIW, and metallurgical processing. For example, Nagorski et al. (2003) conducted a year-long study of four Montana streams and found that

1. Monthly changes were dominated by snow melt and precipitation dynamics, which tend to follow an annual cycle;
2. Daily-scale changes were dominated by episodic events, such as thunderstorms, that resulted in short-term solute and particulate concentrations similar in magnitude to those during the early spring flush; and
3. Diel (i.e., 24-hour) cycles for several constituents and some geochemical parameters were similar in magnitude to much longer (monthly) time scales.

Nimick et al. (2003) reported substantial diel cycles in dissolved metal concentrations (e.g., arsenic, cadmium, manganese, nickel, and zinc) during low-flow conditions along 12 neutral- and alkaline-pH streams draining historical mining areas. In their review of biogeochemical diel cycles in streams, Nimick et al. (2011) stated that the amplitude of diel changes for some water constituents can be as large as changes occurring on annual time scales (see Diel Biogeochemical Processes sidebar). In a study of the seasonality of diel cycles, Nimick et al. (2005) stated that "Diel cycles of dissolved metal concentrations should be assumed to occur at any time of year in any stream with dissolved metals and neutral to alkaline pH." Diel cycles have also been documented in acidic streams and appear to be related to photoreduction of iron oxyhydroxides (e.g., McKnight et al. 1988; Sullivan and Drever 2001; Gammons et al. 2005; Nimick et al. 2011).

Diel variations have serious implications for the collection of water quality data, evaluation of historical water quality data, and collection of baseline water quality data. It is important to consider these short-term variations when designing water quality sampling plans, and it cannot be assumed that a single water sample accurately represents water quality conditions for a given stream. Sampling techniques that effectively address diel variations currently are a topic of research and development. Existing USEPA acute freshwater aquatic-life standards for metals refer to 24-hour average concentrations. As such, it might be useful to consider a sampling program that involves collecting samples throughout the day and night to obtain actual 24-hour average concentrations (see Nimick et al. 2011). Also, consistently collecting periodic samples at a particular site at a specific time of day can minimize the influence of diel variations (Nagorski et al. 2003). However, this approach is problematic for comparisons between sites unless all sites are sampled at the same time under similar hydrologic conditions. An example of misinterpretation

DIEL BIOGEOCHEMICAL PROCESSES IN STREAMS

David A. Nimick, U.S. Geological Survey

The earth's orbit produces the day–night solar photocycle that is part of our daily lives. Many physical, biological, and chemical processes in streams respond to this diel, or 24-hour, cycle and produce persistent cyclic changes in a host of physical and chemical characteristics. The amplitude of some of these diel cycles can be as large as changes observed on annual time scales. The figure presents one example of the multiple diel cycles that can occur in a stream. Changes in streamflow caused by daily changes in the rates of melting snowpack or riparian evapotranspiration, for example, often follow a diel pattern and can be accompanied by changes in water chemistry. Water temperature also increases and decreases on a daily cycle. The most important and ubiquitous diel biogeochemical processes in streams are photosynthesis and respiration by aquatic organisms. This daily life cycle directly affects the dissolved oxygen concentration, pH, carbonate chemistry, and nutrient content of stream water. Because of the diel pH cycle produced by aquatic life and the diel temperature cycle produced by daily solar warming, the concentration of many trace elements also changes on a daily basis because of pH- and temperature-dependent reactions involving photochemistry, adsorption and desorption, and mineral precipitation and dissolution. The possible magnitudes of these changes are summarized in the table below.

Although recognized for more than 150 years, diel biogeochemical cycling in streams was not studied extensively before the 1990s, perhaps because of the expense and inconvenience of the intensive temporal sampling required. Thus, many monitoring and research plans for streams have been developed without sufficient understanding of the potential impact and importance of diel cycling. A recent review by Nimick et al. (2011) summarizes what is known about diel biogeochemical cycling in streams and the changes in aqueous chemistry that are associated with these processes.

Magnitude of diel cycles for dissolved trace elements in near-neutral to alkaline streams. The source of trace elements in these streams is drainage from historical mines, natural weathering of mineral deposits, or geothermal springs.

Trace Element	Maximum Daily Increase, %	Number of Diel Samplings
As	54	>25
Cd	330	12
Cu (pH = 6.8–7)	140	3
Cu (pH > 7)	<10	12
Methyl Hg	93	2
Mn	306	20
Ni	167	1
Rare earth elements	830	2
Se	<10	1
Zn	800	>35

Source: Modified from Nimick et al. 2011.

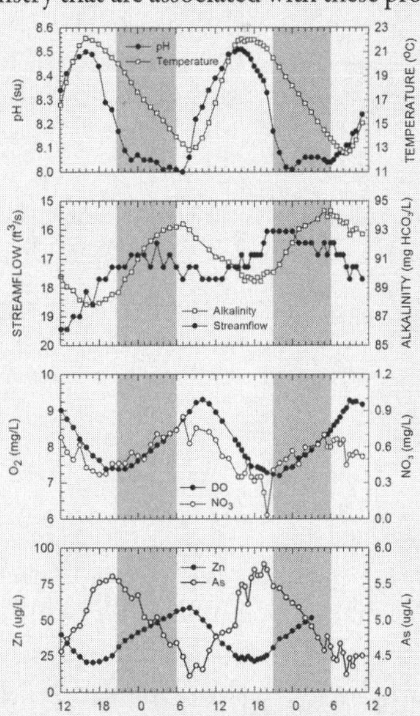

Source: Data from Klein et al. 2003.

Diel variation in physical and chemical characteristics at a U.S. Geological Survey gaging station, Prickly Pear Creek, Montana, July 2001. All constituents shown are dissolved.

SYNOPTIC SAMPLING: A MOVING TARGET?

Christopher H. Gammons, Montana Tech of The University of Montana

Synoptic sampling is often used to quantify changes in solute concentration and load within a watershed. An analysis of spatial changes in solute load reveals likely sources of contamination and informs whether a given solute is conservative or reactive. However, this type of analysis is only valid if short-term (e.g., hourly) temporal changes in solute concentration and load are negligible. To minimize changes in concentration due to snowmelt or rain, synoptic studies are often performed in mid- or late-summer, during baseflow conditions. However, this is also the time of year when diel (24-hour) changes in solute concentration are most pronounced. Failure to consider the effects of diel cycling in a synoptic survey could lead to incorrect conclusions about solute behavior, as shown in the following two examples.

The first example (top figure) is from High Ore Creek, a small, alkaline stream draining abandoned mine lands in Montana. The second example (bottom figure) is from the Mill-Willow Bypass, another small stream in Montana with elevated arsenic concentration. In the High Ore Creek study, one sampler began the day at the bottom of the drainage and worked upstream, collecting samples and streamflow measurements each hour. A second sampler did the same tasks in reverse, working downstream. In the Mill-Willow Bypass study, a pair of samplers spent a day working upstream to collect a set of samples. The next day, nine samplers were dropped off at the same stations and each collected a sample at the same time of day (11:00 a.m.).

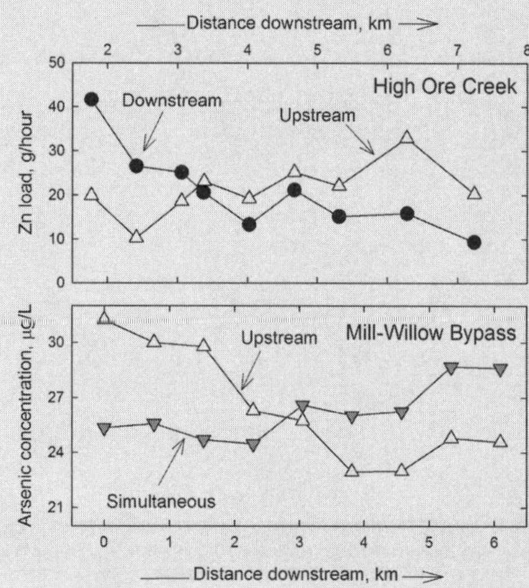

The implications of the different synoptic trends in the examples shown above illustrate the importance of diel cycles. In the case of High Ore Creek, the data set collected by the downstream sampler would indicate continuous attenuation of Zn load along the lower 6 km (3.7 miles) of the stream. In contrast, the upstream sampler would come to the opposite conclusion (i.e., that there are additional sources of Zn load in the stream). Likewise, in the Mill-Willow Bypass example, the upstream data might suggest that arsenic is attenuating in this stream, whereas the simultaneous data lead one to the opposite conclusion. The reason for the different spatial trends could be very perplexing unless the researcher was familiar with diel cycling. In both streams, concentrations of Zn^{2+} decreased during the day at all sampling stations and increased at night, whereas arsenic, an anionic solute, followed the opposite cycle. In both cases, the assumption of negligible short-term temporal changes in solute concentration was incorrect. To assess the magnitude of this problem, it is recommended that a detailed synoptic study should include one or more sets of 24-hour samples. In some cases, the effects of diel cycles will be minimal or absent, which is good to know.

of data due to diel cycles is demonstrated by Gammons et al. (2007) (see Synoptic Sampling sidebar).

3.3.6.2 Suspended Sediment

An important consideration in collecting surface-water samples is whether suspended material is to be included as part of the sample or if suspended sediment is a focus of the investigation. Suspended sediment can be an important pathway for the transport of trace elements due to sorption reactions (Smith 1999). If suspended material is included in the sample, then use of proper sampling devices (e.g., an isokinetic, integrating sampler) is recommended (Edwards and Glysson 1999; Nolan et al. 2005; USGS, variously dated). Acoustic sampling devices are increasingly being used to determine suspended sediment loads when chemical data are not required. Horowitz (2008) discusses issues involved in suspended sediment sampling.

3.3.6.3 Discharge Measurements

Stream-water and treated-water flow/discharge measurements can be invaluable when comparing solute concentrations between different sampling events because concentration variations and flow variations are generally linked. Discharge measurements are also a typical requirement for permitting. In mined and mineralized areas, it is common to see relationships between solute concentrations and flow. Predominant causes of changes in hydrologic conditions include evapotranspiration, snow melt, thunderstorms, discharge from treated water, and loss or gain of flow due to infiltration through the stream bed. In their evaluation of historical surface-water quality data for the Red River Valley, New Mexico, Maest et al. (2004) found that the time of year has a major effect on water quality due to changes in hydrologic conditions. They also reported that episodic events, such as thunderstorms, cause flushing of sulfide oxidation products from altered areas, and that alkalinity values decrease during storm events, which indicates that there can be short-term changes in buffering capacity. Johnson and Thornton (1987) and Jambor et al. (2000) also reported elevated metal concentrations in streams draining mined areas due to flushing of stored acid and metals from soils and mining wastes by melting snow and rainfall events. McKnight and Bencala (1990) and Boyer et al. (1997) reported flushing of dissolved organic carbon (DOC) and other constituents from soil during spring snow melt. Therefore, when collecting stream-water samples, it is important to know where stream-water samples fall on the hydrograph. It is good practice to obtain samples from a given stream under a variety of hydrologic conditions.

Streamflow information is critical to computing the load of water constituents, predicting floods, and designing engineering structures and mitigation strategies at mine and processing sites. Load is the product of concentration and stream discharge and hence represents the quantity of an analyte. Depending on the question to be answered or the application, analytes can be reported as concentrations or as loads. Regulatory criteria and toxicological data are generally reported as concentrations. Loads can be used to address questions that involve accumulated mass or volume and treatment efficiency. TMDLs, which are the amount of contaminant that an impaired water body can receive and still meet water quality standards (Walton-Day et al. 2012), are an example of the application of loads.

Many techniques can be used to obtain stream-water flow or discharge measurements (e.g., Rantz et al. 1982). USEPA (2001c) provides guidance on performing flow measurements at mine sites. Turnipseed and Sauer (2010) and Kennedy (1984) provide detailed descriptions of discharge measurements, equipment, and calculations. Hydroacoustic meters (e.g., acoustic Doppler current profilers [ADCPs] and acoustic Doppler velocimeters) represent a relatively new approach for determining streamflow. The ADCP is rapidly replacing traditional mechanical

WHERE'S THE LOAD?

Briant A. Kimball, U.S. Geological Survey

The total maximum daily load (TMDL) process requires an accurate accounting of loading along the course of a stream or river. When a load is determined at the outlet of a catchment (see figure), the loading of individual sources within the catchment is integrated into one sample; no spatial detail is available to compare and quantify loads from the various sources. An approach that combines two methods, synoptic sampling and tracer-dilution discharge, has been used to provide the kind of spatial detail that helps to compare loads for a TMDL process (Walton-Day et al. 2012).

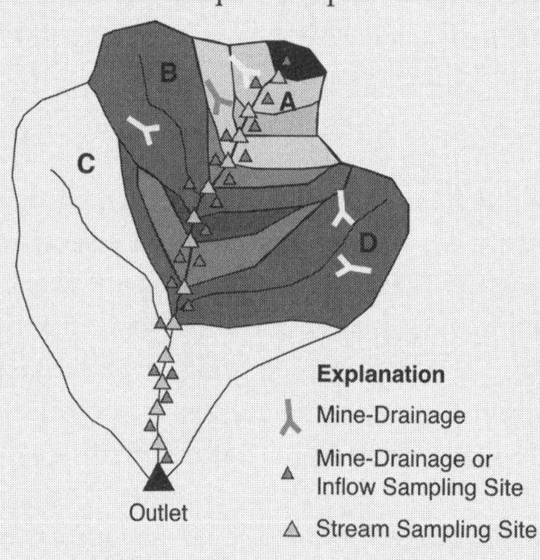

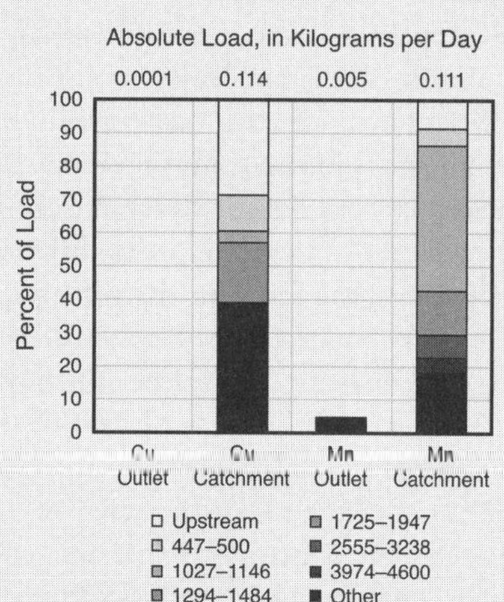

Synoptic sampling within a catchment is accomplished by walking the stream and identifying inflows to the stream. Stream sampling sites are located upstream and downstream from these inflows, or from representative inflows if there are numerous inflows. Stream sites are also located upstream and downstream from areas of likely groundwater or dispersed inflow. In any case, sufficient stream-sampling sites are selected so that spatial detail on the order of 100–300 m (328–984 ft) is achieved. Thus, as suggested in this figure, detailed synoptic sampling divides the catchment into *slices* that each account for a portion of the total load. The same spatial detail is available for measurement of discharge using the continuous tracer injection method described herein (see Appendix 3). With the combined concentration and discharge data, loads can be calculated at each of the stream sites along the catchment study reach (Kimball et al. 2002, 2007). Each slice of the catchment can be compared in terms of constituent loading rather than concentration, providing the frame of reference that is needed for a TMDL evaluation.

Results from a catchment that is affected by several areas of mine drainage are illustrated in the bar graph and compared to the load that would have been measured at the outlet of the catchment. The absolute quantity of the loading, in kilograms per day, is given at the top of each bar. The portion of the load that can be attributed to specific sections, or slices, of the catchment is indicated by the different shades. Those stream segments that contribute the most copper load are not always the same as those that contributed the greatest load of manganese. The load that would have been observed only by a sample at the outlet was negligible for copper and only slightly greater for manganese. This shows the substantial difference in accounting by taking an outlet perspective rather than a detailed catchment assessment. The difference can be attributed to attenuation of loads along the course of the stream as metals are removed by physical and biogeochemical processes. However, almost all of the load that is removed by attenuation is subsequently flushed by snow-melt runoff and still contributes to the loading on an annual basis. This seasonal variation of load from the catchment should be quantified by sampling over time at the outlet, and the slices of the total should be quantified by a detailed catchment approach.

current meters for streamflow measurements (Muste et al. 2007; Oberg and Mueller 2007). Treated water discharges are generally monitored using magnetic flowmeters.

Use of tracers in surface water is an effective approach to determine gain or loss of water quantity. An approach using a tracer-dilution method to quantify stream-water discharge and synoptic sampling to quantify solute concentrations has been developed to determine metal loading in streams located in mined and mineralized areas (see Where's the Load sidebar, Appendix 3, and case study by Walton-Day et al. in Appendix 5; Kilpatrick and Cobb 1985; Kimball 1997; Kimball et al. 2007, 2010). Using this approach, it is possible to identify and quantify the principal inflows of metals, the presence of nonpoint source, dispersed inflows of metals, and the amount of natural attenuation of metals in a stream. Tracer-dilution methods can also be used together with other tools, such as isotopic measurements, to identify sources and attenuation processes (e.g., Borrok et al. 2009).

3.3.6.4 Modeling Considerations

Geochemical, toxicological, and hydrogeochemical modeling requirements need to be considered when preparing a sampling and analysis plan. Modeling hydrogeochemical processes using a variety of numerical models is an expanding area of research and development. Both numerical and water mass-balance models can be used to refine conceptual models prior to mine startup, evaluate treatment and mitigation options, manage water on the site during operation, and guide selection of appropriate reclamation options during mine operation and closure. Voss and Letient (2006) and Eary et al. (2009) have demonstrated how existing numerical models can be reconfigured and used as decision support tools at mine sites.

For geochemical modeling, it is necessary to have complete dissolved water analyses including major, minor, and trace elements, and both anions and cations. Additional important determinations include pH, temperature, iron speciation, and direct measurement of species for elements of concern. Also, specific conductance, alkalinity/acidity, DO, redox conditions, and total dissolved solids can provide useful information. Current analytical techniques offer simultaneous analysis of a number of chemical elements; therefore, it is relatively simple to obtain complete water analyses. Also, complete water analyses can serve as a guide to unanticipated environmental concerns or to potential by-products that may be recovered. Nordstrom and Munoz (2006) discuss information needs for various applications of geochemical modeling. For transport modeling, it is necessary to have water-flow or discharge measurements. For inverse modeling, it is necessary to have mineralogical information about the geologic material in contact with the water. Caruso et al. (2008) discuss recent advances in metals fate and transport modeling. Geochemical modeling is the topic of another handbook in this series (see Nordstrom et al., forthcoming) and is also discussed in Alpers and Nordstrom (1999) and Nordstrom (2004). Butler (Appendix 5 of this handbook) provides a case study related to sampling and geochemical modeling.

Toxicological considerations are also an important aspect of water quality studies. Stillings et al. (Appendix 5, this handbook) provide a case study about selenium cycling and bioaccumulation at Idaho phosphate mines. Toxicological modeling of water quality is beginning to have a presence in federal regulations. For example, USEPA incorporated the biotic ligand model (BLM) into its recently revised aquatic life ambient freshwater quality criteria for copper (USEPA 2007a). The BLM is a computational model that predicts metal toxicity to biota in aqueous systems for one metal/organism pair at a time (Di Toro et al. 2001; Santore et al. 2001; Appendix 3). Required input for the BLM includes temperature, pH, DOC, alkalinity, and dissolved concentrations for calcium, magnesium, sodium, potassium, sulfate, and chloride. Although DOC samples are rarely collected, DOC is a required input parameter for the BLM and can have a significant influence on predicted toxicity of some metals such as copper (e.g., Smith et al. 2006, 2009). Therefore, it is important to include DOC in surface-water sampling plans. A research version of the BLM (see Section 3.13) includes computations for copper, silver, cadmium, and zinc. In addition to determining site-specific water quality criteria, the BLM can be used for additional applications such as determining the time of year when the water chemistry is most toxic, which is often difficult to determine without a rigorous and expensive toxicity-testing program.

3.3.7 Groundwater Samples

There are many reasons for sampling groundwater, such as tracking a contaminant plume, determining the existence or extent of contamination, compliance monitoring, or performing water-supply measurements. Groundwater sampling can be conducted during any of the phases of mining depending on the objectives of the investigation. Preserving the integrity of groundwater samples is not an easy task. LeGrand and Rosen (1998, 2000) called for a widespread change in mindset for groundwater contamination investigations. They listed three problematic philosophies and procedures used in groundwater contamination studies, which include (1) almost exclusive reliance on quantitative, mathematical, precision-oriented data; (2) an incomplete or belated conceptual model of hydrogeologic characteristics; and (3) insufficient use of existing databases and other available information. They encouraged developing an early *prior conceptual model explanation* that is largely derived from existing information to conduct a cost–benefit evaluation of necessary groundwater protective actions.

General Web sites for online resources related to groundwater are listed in Section 3.13. Several methods for groundwater sampling are documented by ASTM International and USGS (variously dated). Some general groundwater references include Freeze and Cherry (1979), Barcelona et al. (1985), USEPA (1991), Chapelle (1993), Fetter (2000), Alley et al. (2002), Appelo and Postma (2005), Nielsen (2006), Nielsen and Nielsen (2007), Younger (2007), and ASTM (2007).

Significant advances in groundwater sampling include addressing colloid formation and sampling (e.g., McCarthy and Zachara 1989; Backhus et al. 1993; Gschwend and Reynolds 1987; Kearl et al. 1992), low-flow sampling, and aquifer heterogeneity with emphasis on small-scale differences. The potential of hydrochemical communication between aquifers is another important consideration. As with other sampling media, it is critical to determine the sampling objectives and data applications prior to locating and designing groundwater wells and sampling groundwater. An important question for groundwater sampling is if it is necessary to focus on a particular portion of the formation because information about small-scale spatial differences in groundwater quality requires planning and specialized sampling techniques. Locating wells with respect to site geology and hydrology are also important considerations. Surface geophysical techniques can provide information to aid in properly locating well sites. Some geophysical techniques can be used to map subsurface plumes and locate sources of acidity (see discussion of geophysical methods in Appendix 3).

Filtration is a controversial issue for groundwater samples (e.g., Puls and Barcelona 1989; Puls et al. 1990). Puls and Powell (1992) have reported that colloids as large as 2 µm can be mobile in porous media. Federal and state regulations vary on this issue, and some analytical procedures require that a sample be filtered. If filtration is used, it is preferable to filter in the field using in-line filtration, especially for unstable samples (e.g., anoxic groundwater). It is good practice to collect unfiltered samples in addition to filtered samples.

Tracers and isotopes in groundwater are generally used to characterize groundwater flow and to identify potential contaminant pathways (e.g., Wolkersdorfer 2002; Manning et al. 2008;

Walton-Day and Poeter 2009; see Appendix 3). An overview of tracers is given by Healy and Scanlon (2010).

3.3.7.1 Representative Groundwater Samples

It is very difficult to obtain a groundwater sample that is representative of the formation water. Several researchers have addressed the difficulty of obtaining representative groundwater samples (e.g., Parker 1994; Puls et al. 1992). Nielsen and Nielsen (2007) list the following factors that influence the representative nature of groundwater samples:

- Formation and well hydraulics
- Groundwater flow paths
- Flow-through wells and well screens
- Hydraulics within a well between sampling events
- Sampling point placement and well design, installation, and maintenance
- Placement of sampling point with respect to contamination source(s)
- Placement of sampling point intake in the zone of highest hydraulic conductivity
- Drilling and installation methods
- Suitability of design with regard to material selection, diameter, depth, screen length, and screen slot size
- Construction methods including placement of annular seal materials and filter pack materials
- Method, timing, and duration of sampling point development
- Long-term maintenance of the sampling point (e.g., compromises in well integrity)

3.3.7.2 Well Design and Purging Methods

Proper well design is an important consideration for groundwater sampling. The objectives of the investigation should dictate the well's design. For example, it is generally preferable that the screen interval only encompass the sampling zone of interest so that mixing of different water sources does not occur in the well bore (e.g., Church and Granato 1996). Nested wells can be used to allow sampling of multiple depths at the same location.

Purging and collection methods have a significant impact on the sample quality and integrity. Many traditional groundwater sampling procedures were designed for water supply investigations and are not suitable for contaminant-related investigations. Depending on the study objectives, it can be helpful to probe the water column before pumping to identify the presence and degree of stratification, and to geochemically characterize different zones. Traditional purging procedures require that three to five well-casing volumes be purged to remove stagnant water from the well prior to sampling. This purging is frequently done using high-speed pumping rates or with a bailer. This practice can be problematic for water quality sampling because rapid removal of water can lead to increased turbidity (producing particulates in the water sample; Kearl et al. 1994), disruption of well packing materials (producing contamination), and drawing water into the well that is not representative of the formation water adjacent to the well. Puls and Powell (1992) have reported that sample-collection practices that induce artificially high levels of turbidity have the greatest negative impact on sample quality. Filtration should never be considered as a remedy for improperly collected groundwater samples. Fracture porosity in the formation can also be very important because extensive pumping can draw water from a large volume of the formation. For example, McCarthy and Shevenell (1998) reported that pumping 10 L of water from a well in rock with fracture porosity could conceivably draw water from a volume of 100 m^3 of rock.

Recent reports suggest that it is important to keep subsurface disturbance and sampling handling to a minimum when collecting groundwater quality samples. Hence, it is important to use appropriate purging and sampling equipment when collecting groundwater samples. Low-flow (low-stress), minimum drawdown purging and sampling techniques are being recommended for most groundwater quality sampling (Puls and Barcelona 1996; USGS, variously dated; Ritchey 2002; Yeskis and Zavala 2002; ASTM 2005). Although time-consuming, these techniques reduce the need to dispose of large volumes of purged water, minimize sampling artifacts, reduce stress on the formation, and result in better consistency of groundwater samples (e.g., Sevee et al. 2000). The flow rate is site specific and this approach strives to cause minimal or no drawdown in the formation adjacent to the well. During the purging process, several indicator parameters (e.g., pH, temperature, specific conductance, turbidity, DO, redox potential) need to be monitored to determine when they stabilize (e.g., Wilkin et al. 2001; see USGS, variously dated, for parameter stabilization and calibration guidelines). Barcelona et al. (2005) have concluded that stabilization of field water-quality parameters is an effective indicator of when to collect samples. If possible, it is best to use dedicated downhole sampling devices.

3.4 SOLID SAMPLING CONSIDERATIONS

The purpose of collecting solid samples at mine and processing sites (other than for exploration purposes) is normally to quantify metal content at the sampling location. Additional reasons can include establishing baseline conditions, examining variation in lithologies, assessing the potential for acid generation, assessing the potential for contaminant release, locating contaminant source(s), providing input for modeling efforts, or meeting regulatory requirements. Samples collected for metallurgical purposes and assays can often be useful for environmental purposes.

3.4.1 Mining- and Metallurgy-Related Samples

When sampling to determine potential for acid generation or contaminant release, it is good practice to classify solids into geological categories (e.g., soil, overburden, waste rock, ore, sub-ore, tailings, grab, chip, composite; see Appendix 3 for discussion of various solid materials). Environment Canada (2009) outlines site-specific programs for predicting wastewater quality throughout the mine life cycle, and Miranda et al. (2005) discuss the importance of minimizing ARD and suggest that testing for ARD potential be done as a part of an environmental impact assessment in advance of mine development.

Runnells et al. (1997) provide a methodology for sampling mine waste rock and process tailings. They state, "Briefly, the method is based on the use of a statistical approach to determine, illustrate, and defend the adequacy of the sampling. [We do] not believe that there is a *correct* number of samples for characterizing a facility. That is, there is no general rule that can (or should) be followed, such as a given number of samples per ton of tailings, per acre of impoundment, or per foot of drillcore. Each facility is different, and the adequacy of sampling must be tailored to the facility." Price (2009) provides guidance about geological characterization at mine sites and the use of block models for environmental applications. He states, "The recommendation here and previously is that the final sampling frequency be determined site specifically based on the variability of analysis results for critical parameters, prediction objectives and required accuracy" (Price 2009). Hence, site-specific conditions are critical in determining what and how much to sample at mining sites (see the following three sidebars about mine rock sampling). McLemore provides a case study that addresses mine-rock sampling (Appendix 5, this handbook).

WASTE ROCK SAMPLING

Patsy B. Moran, Arcadis U.S., Inc.

Waste rock geochemical characterization is primarily conducted to assess the potential for acid rock drainage and metal leaching as a result of excavation and subsequent exposure to oxygen and water. The objectives of geochemical characterization will change as mining projects advance from exploration to postclosure. Early-stage characterization is often carried out to assist with design of facilities and development of management plans. Operational programs frequently focus on meeting permit requirements, and postclosure characterization may be needed if drainage chemistry deviates from predictions. Regardless of the project phase, selection of representative samples plays a critical role in the effectiveness of any characterization program.

Although waste rock sample selection should be conducted on a site-specific basis due to inherent site and project variability, several factors should be considered for each project, including selection of

- Discrete samples from the primary rock types based on geologic or other operationally defined units,
- Geospatially distributed samples from throughout the area to be mined, and
- Enough samples to adequately characterize the range of acid–base characteristics for each rock and alteration type with focus on those most likely to be problematic.

Arguably, the most difficult of these items to address is determining how many samples will be needed to meet project objectives. Some important considerations include

- Heterogeneity of the rock types,
- Quantity of disturbed rock (e.g., tons, wall rock area),
- Desired level of confidence, and
- Project phase.

Several recent, publically available documents have detailed guidance for properly conducting waste rock sampling:

- *Prediction Manual for Drainage Chemistry from Sulphidic Geologic Materials* (Price 2009)
- The Web-based *Global Acid Rock Drainage Guide* (GARD Guide; INAP 2009)

Both documents also provide in-depth information to assist with development of waste-rock characterization programs and drainage-chemistry predictions.

The mine geologic model and block model are powerful tools that may be used for several purposes, including to identify target populations, assign environmental properties to various units, and discretize waste rock temporally and spatially that can inform material handling plans based on geochemical characteristics. Modern mine planning software packages (e.g., MineSight [Mintec, Tucson, Arizona], GEOVIA [Dassault Systèmes, Vancouver, British Columbia, Canada], Vulcan [Maptek Pty Ltd, Adelaide, Australia]) allow detailed three-dimensional definition of not only ore properties, but also geochemical data that can be applied to mine-waste management during operations and for waste rock placement as part of operation and closure plans. The excavation of the entire deposit, all ore and waste rock, can be simulated and waste-rock placement plans can be evaluated in advance of actual production to ascertain various waste-rock handling and placement strategies.

SAMPLING SURFICIAL MATERIAL FROM MINING WASTE-ROCK PILES

Kathleen S. Smith, U.S. Geological Survey

Site characterization of mining wastes requires collection of representative samples of the mine-waste material. This is generally difficult because of the compositional and spatial heterogeneity of the waste material. A statistically based strategy was developed for sampling the surficial material of mine-waste material for use in screening and prioritizing historical waste piles (Smith et al. 2000).

The target population is defined as the upper 15 cm of a waste-rock pile. To minimize sampling errors, a composite sample of at least 30 increments (subsamples) was collected, which reduces the grouping and segregation error (GSE). To minimize the fundamental sampling error (FSE), the sample is sieved to a <2 mm size fraction. For an FSE below 2%, it is necessary to collect at least 1 kg for the final sieved composite sample. This combination of maximum particle size and 2% fundamental error keeps the total sample mass to a weight that can be easily carried in a backpack after sample collection. Samples are collected with a stainless-steel trowel, and each increment is successively placed in a plastic bucket. If samples are collected wet, they are air-dried and efforts made to break up clods prior to sieving.

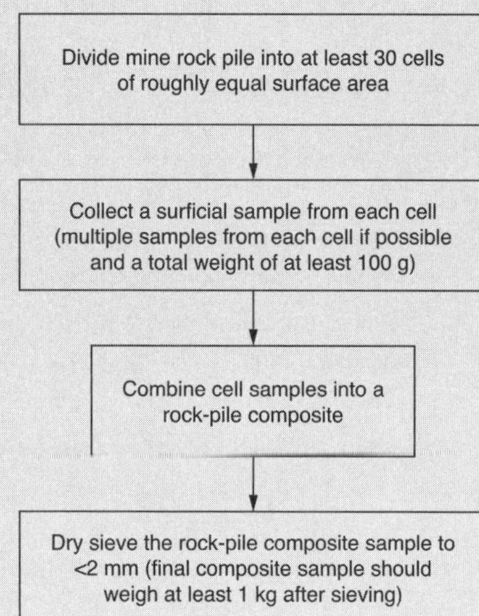

In terms of average properties, one 30-increment composite sample collected using this sampling strategy contains as much information relative to average value as 30 individual grab samples at $\frac{1}{30}$ of the analytical cost. This sampling strategy has been used in conjunction with the U.S. Geological Survey field leach test (Hageman 2007) to characterize mine-waste material (e.g., Hageman and Briggs 2000) and to provide a comparison between the leaching characteristics of waste-rock piles and naturally altered areas (Smith et al. 2007). It can also be used for other target populations, such as lithologic units and soils, and has been used to collect soil, sediment, and ash samples from natural disasters (e.g., Plumlee 2009).

It is advisable to sample all lithologies and to determine the degree and range of alteration, acid-generating potential, mineralogy, and bulk chemistry (major, minor, and trace elements). Two big issues are common in sampling mine-waste facilities: (1) the typically very heterogeneous nature (both compositionally and grain size), which makes it difficult to obtain a representative sample, and (2) homogenization and subsampling of the material during sample preparation for analysis. These issues are discussed in more detail in Sections 3.4.8 and 3.4.9, and in some of the case studies in Appendix 5. Bulk chemistry is important for ore and waste materials, ore stockpiles, ultimate pit surfaces, and amendment materials blended with wastes because it allows identification of potentially problematic constituents and pinpoints which zone(s) of materials require further testing. Acid–base accounting (ABA) methods and kinetic testing are discussed in the prediction volume of this series (Williams and Diehl 2014) and will not be addressed here.

A QUICK AND DIRTY ASSESSMENT

Thomas R. Wildeman, Colorado School of Mines

In Colorado there are thousands of orphaned mine-waste piles. The question is, how can one assess these piles in a cost-efficient manner to find out which are the most harmful to the environment? For more than a decade, scientists at the U.S. Geological Survey (USGS) and the Colorado School of Mines have been considering this problem and have developed a fairly simple scheme to assess potential impacts from a waste pile (Ranville et al. 2006; Wildeman et al. 2007). Two features are at the heart of the scheme. The first is the development of a simple method for obtaining a multi-increment sample that is based on sound sampling procedures and statistical rigor (Smith et al. 2000). The second feature is the development of the mine waste decision tree (MWDT), shown herein.

Mine Waste Decision Tree

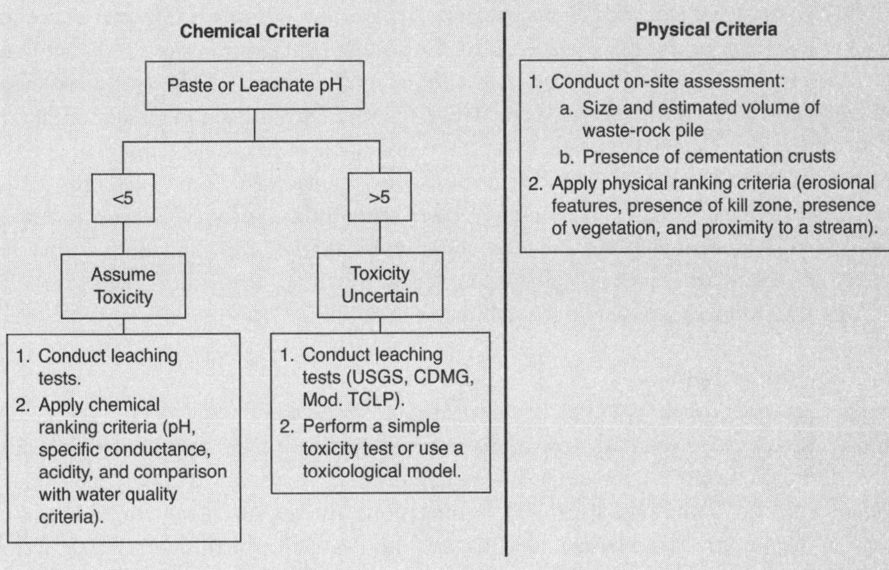

Source: Adapted from Ranville et al. 2006 and Wildeman et al. 2007.

The MWDT involves three interesting features. First, there is a separate chemical and physical assessment. The chemical criteria rate the availability of contaminants, and the physical criteria rate the ability to deliver the contaminants. The second feature is that if the paste pH of the composite sample taken from the pile is below 5, then the decision is to assume toxicity to the aquatic environment. This assumption is based on geochemical principles and on experiences of sampling and assessing hundreds of waste piles in the Rocky Mountain region. The third feature is three simple leaching tests that have been developed to assess the actual contaminants: The USGS field leach test (FLT; Hageman 2007), the Colorado Division of Minerals and Geology (CDMG) test, and a modified toxicity characteristic leaching procedure (Mod. TCLP) test. These tests are modified versions of standard regulatory tests, so they are primarily suitable for assessment projects. However, they are simple and inexpensive, and the FLT and CDMG tests can be performed on-site. The procedures for securing a composite sample and the leaching tests are robust enough that they can be used whenever a quick and dirty assessment needs to be done. They also have been used effectively to assess the rubble from the 9/11 disaster and the sediment left from the flooding caused by Hurricane Katrina.

3.4.2 Soil Samples

Numerous references involve soil characterization and soil sampling (e.g., Mason 1992; USEPA 1995; Sparks 2003; Carter and Gregorich 2008). The Soil Science Society of America published a five-volume series titled Methods of Soil Analysis that covers physical and mineralogical methods, microbiological and biochemical properties, and chemical methods. USEPA considers soil to be an exposure pathway rather than a migration pathway because contact between contaminated soil and a receptor is initiated by the receptor (USEPA 1995). USEPA soil screening guidance (see Section 3.13) provides a framework for developing risk-based, soil screening levels (SSLs) for protection of human health. The SSLs do not represent national cleanup levels; they are intended to be used by site managers to eliminate areas, pathways, and/or chemicals of concern at National Priorities List sites.

Soil consists of heterogeneous mixtures of solid particles (inorganic and organic), microorganisms, air, and water. Because soil is so chemically, physically, and biologically complex, it is important to carefully define the target population and to ensure that it is consistent with the sampling objectives and methods. Soil samples need to be dried immediately after collection at ambient or low heat, and loosely packed into the sample container. For samples that contain jarosite, Desborough et al. (2010) report that the common practice of heating above approximately 100°C (212°F) during *drying* can change the structural nature of jarosite minerals. In situations where there are hydrous sulfate salts, it is preferable to freeze the samples or cores.

Soil sampling and analysis techniques can be applied to investigating the weathering, erosion, and stability of mine waste rock piles, and several of these techniques are described in Appendices 2 and 3. McLemore et al. (2009a) discuss results from a series of investigations at the Questa mine in New Mexico, where they employed a variety of physical, geochemical, petrographic, and mineralogical techniques to evaluate the stability of rock piles.

3.4.3 Sediment Samples

Many publications provide guidance for collection of sediment samples (e.g., Truckenbrodt and Einax 1995; Edwards and Glysson 1999) and suspended sediment (e.g., Nolan et al. 2005). Brandvold and McLemore (Appendix 5, this handbook) provide a case study that describes aspects of sediment sampling adjacent to mining sites. The approach for sampling sediment depends on the objectives and elements of concern. Studies of the distribution of trace elements in relation to size fraction generally show that most trace metals (e.g., copper, iron, manganese, and zinc) are concentrated in the finest fractions of the sediment. Typically the <150-µm size fraction is collected (by wet sieving) and analyzed. However, the possibility that some elements of interest can be preferentially bound to coarser, sandy material in the vicinity of a mine site also needs to be considered and investigated. Vertical profiles can provide useful information about metal residence and mobility (e.g., Church et al. 2007a). In collecting sediment samples for acid-volatile sulfide/simultaneously extracted metals (AVS/SEM; see Appendix 3) or other solid samples from reducing environments, care should be taken to protect samples from oxidation by enclosing them in sealed containers (under nitrogen if possible) or, better yet, to collect them under nitrogen followed by freezing. Equilibrium partitioning sediment benchmarks for metals are sediment-quality guidelines developed by USEPA that relate chemical concentrations in sediment to potential biological effects on benthic organisms (USEPA 2005; see Appendix 3 and case study in Appendix 5 by Seal et al.).

If sediment samples are not collected from reducing environments or being used to determine AVS/SEM, then wet samples need to be immediately air-dried, or dried at temperatures less than 40°C (104°F), to avoid changing the clay, iron, and sulfate mineralogy.

3.4.4 Bulk Composition and Geoavailability

Bulk composition of solid material can be analyzed in the field by portable X-ray fluorescence (PXRF) or by standard wet chemical methods (see Appendix 3). The PXRF apparatus offers rapid in situ semiquantitative analysis; wet chemical analysis offers potentially greater accuracy but is a relatively slow and more expensive process. PXRF allows detailed mapping of contamination hot spots (based on bulk composition), and the perceived relative lack of accuracy is outweighed by the large number and wide range of samples that can be analyzed. Field leaching tests, such as the USGS field leach test (Hageman 2007), coupled with field test kits can be used to ascertain readily soluble (i.e., geoavailable; see Smith and Huyck 1999; Smith 2007) constituents at the field site (see Appendix 3 and Figure 4.4).

The transport and fate of trace metals are generally controlled by their partitioning between aqueous and solid phases. Understanding the partitioning mechanisms of these metals is critical for evaluating and predicting their mobility and bioavailability (Smith 2007). Organic constituents, iron and aluminum oxyhydroxides, and manganese oxides are generally the most important sinks for trace metals due to sorption reactions (Smith 1999). Generally, these phases occur as coatings on other mineral surfaces, which makes a molecular level of understanding difficult (e.g., Diehl et al. 2006).

The importance of mineral surfaces to reactivity and mobility of metals in the environment is increasingly better understood (Sparks 2005; Riviere and Myhra 2009). Fundamental understanding of minerals and their reactions in natural systems requires examination at the molecular and atomic scale. Hochella (1995) provides a review of some surface characterization techniques, and Fenter et al. (2002) provide an overview of the theory and applications of synchrotron radiation to the analysis of soil surfaces. Surface analytical techniques include a variety of spectroscopic methods that have been used in geological and environmental applications to examine the solid–water interface and surface speciation of elements. X-ray absorption spectroscopy (XAS; Kelly et al. 2008), including X-ray absorption near-edge structure (XANES; e.g., Foster et al. 1998; Fleet 2005; Walker et al. 2005) and extended X-ray absorption fine structure (EXAFS; e.g., Kim et al. 2004), is valuable because it can provide information about a particular element and its speciation and oxidation state, which is particularly useful for amorphous materials and metal-sorption studies (the techniques are discussed in more detail in Appendix 3). More specialized methods are being developed, and applications to earth materials will continue to grow.

3.4.5 Mineralogical and Petrographic Characterization

Mineralogical characterization can provide fundamental information regarding the potential for acid generation, neutralization, and contaminant release, yet it has historically been neglected. Mineralogy and texture are key factors that influence acid generation, neutralization, and metal release (e.g., Kwong 1993; Cabri and Vaughan 1998; Plumlee et al. 1999; Vaughan and Wogelius 2000; Munroe et al. 2000; Diehl et al. 2006; Shaw and Mills 2012). Kirk (Appendix 5) provides a case study that describes the use of historical mineralogical data combined with laboratory- and field-scale monitoring data to predict postmining water quality. Mineralogical studies are performed to determine primary and secondary mineralogy, alteration information, and identification of sulfide minerals, carbonate minerals, and other acid-neutralizing minerals (e.g., biotite; e.g., Neubert 2000; Hammarstrom and Smith 2002). Pitard (2010) recommends a thorough mineralogical or microscopic study at the early stage of a sampling project because it is important to understand how a given trace constituent behaves in the material to be sampled. In particular, textural information such as degree of particle/phase liberation, phase association, and particle size and morphology can provide important information about the potential behavior of

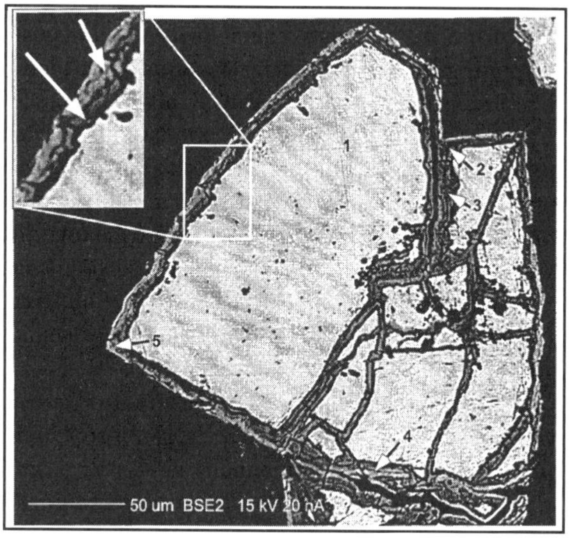

Source: Munroe et al. 2000.

FIGURE 3.3 Backscattered electron microprobe image of an oxidized pyrite grain with Fe-hydroxide and Fe-sulfate alteration rinds of low and high atomic mean number (sample is from Laramide polymetallic veins). Brecciated parts of this grain are interpreted to have undergone hydrothermal fracturing and subsequent alteration along fractures and cementation by iron oxides.

constituents. A variety of techniques can be used for mineralogical, compositional, morphological, and textural characterization. Some traditional techniques include optical microscopy (e.g., Kerr 1977; Craig and Vaughan 1994; Nesse 2004), x-ray diffraction (XRD; e.g., Nuffield 1966; Bish and Post 1989), and electron microbeam methods (e.g., White 1985), including scanning electron microscopy (SEM) and electron-probe microanalysis (EPMA; e.g., Reed 1996; McGee and Klaus 2001; see Appendix 3).

Petrographic analysis provides information on the texture of the material and the relationship of the minerals to each other. Petrographic analysis of rocks has traditionally been performed with reflected and transmitted light microscopy using thin sections and point counting to provide a modal mineralogy, percent porosity, and grain size (van der Plas and Tobi 1965; Delvigne 1998; McLemore et al. 2009b). However, this method typically does not provide accurate and reproducible mineral proportions for sedimentary, volcanic, and soil-like material because lithic fragments, groundmass, and matrix are typically identified as separate phases, not as specific minerals. In addition, groundmass phases in volcanic rocks can be too fine grained to identify individually, and making thin sections of intact soil-like material is difficult because much of the material is not well cemented. Petrographic analysis is also useful to determine residual sulfide minerals and residence phases of trace elements in slags (e.g., Piatak et al. 2004; see case study by Piatak et al. in Appendix 5).

Optical microscopy, EPMA, and SEM can provide valuable information about the morphology and texture of the material. For example, microanalytical studies can determine the morphology of pyrite (whether it is chemically zoned, coarse grained, fine grained, or framboidal) and thereby identify forms that are more accessible to oxidation. They can also determine if pyrite is encased by gangue minerals, such as quartz or clay, and consequently less accessible to oxidation. Iron oxidation rims surrounding minerals can protect the mineral grains from subsequent oxidation (see Figure 3.3; Munroe 1999; Munroe et al. 1999, 2000). Spot analysis

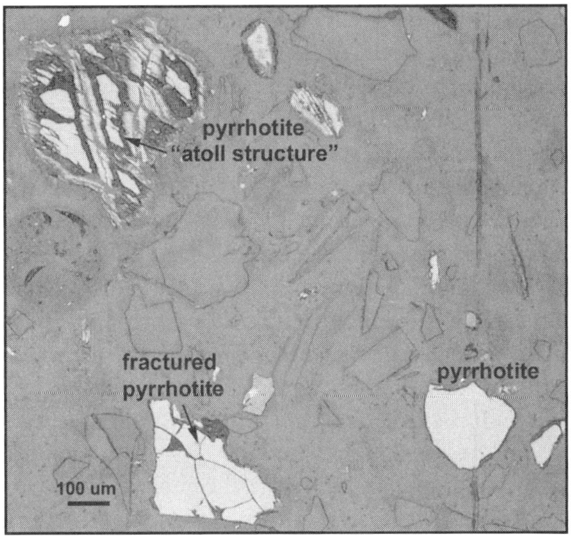

Courtesy of Sharon F. Diehl, U.S. Geological Survey.

FIGURE 3.4 Backscattered electron image of pyrrhotite grains in a tailings deposit, Elizabeth mine, Vermont. The pyrrhotite grains have undergone varying degrees of weathering, which could be due to trace-element content or differing lengths of exposure in the tailings. Atoll structure develops through the partial dissolution of the sulfide grain (Jambor 2003) and alteration to secondary iron minerals, which leaves residual islands of pyrrhotite.

and element-intensity maps by SEM or EPMA identify trace elements associated with the main mineral phase and characterize their residence (i.e., whether metal impurities are homogeneously distributed within the mineral or are associated with mineral inclusions). This information can be very useful in identifying the source of contaminants in water samples (e.g., Diehl et al. 2006, 2008). Incorporating mineralogical characterization into predictive testing and environmental impact assessments can help define potential environmental challenges, which allows for better ore stockpile and waste repository design as well as estimates of remediation and mitigation costs, if necessary.

Sulfide minerals such as pyrite, marcasite, and pyrrhotite, which are the prime generators of ARD, are commonly hosts to elements of concern such as arsenic, cobalt, copper, nickel, lead, selenium, or zinc. Trace metals change the physical properties of minerals, commonly increasing reactivity and solubility by lattice distortion during element substitution. Figure 3.4 is a backscattered electron image of pyrrhotite grains in a tailings deposit from the Elizabeth mine in Vermont. The important observations to note are that the pyrrhotite grains show varying degrees of weathering, ranging from fairly pristine grains to fractured grains, and that fractures allow ingress of oxygen and fluid that partially dissolve pyrrhotite grains, which results in alteration of the rims to marcasite and iron-oxide minerals. These sulfide minerals will continue to slowly weather, and they remain in the tailings deposit as a repository for future acid drainage (AD) and metal release.

It is important to quantify mineral phases that influence the composition of effluents from drainage of ore and waste materials. Not all sulfide minerals have the potential to produce acidic drainage, and iron-sulfide minerals may not produce acidic drainage in some materials. For example, galena, sphalerite, covellite, bornite, and chalcocite do not release acid when oxidized (Plumlee et al. 1999). Conversely, minerals other than sulfides can generate acidic drainage. For example, some forms of jarosite (e.g., Lapakko and Berndt 2003; Desborough et al. 2010) and soluble sulfate salts (e.g., Cravotta 1994; Jambor et al. 2000) release acidity when in contact with water. Identifying neutralizing minerals, such as carbonates and clay minerals, is an important step in assessing the neutralizing potential of the material (e.g., Jambor et al. 2002, 2007).

Determining quantitative mineral abundance by whole-rock XRD analysis is occasionally difficult because of the effects of mineral crystallinity (i.e., acidic waste materials generally consist of a high proportion of amorphous material), preferential orientation in the sample mount, differential absorption of X-rays, and overlapping peaks by different minerals, which all affect the diffractogram patterns and make their interpretation difficult. Some minerals, such as hematite and other iron oxides, have poor crystallinity and are not always detected by XRD. The Rietveld method attempts to overcome some of these issues (Bish and Howard 1988; Bish and Post 1993; Young 1993; Oerter et al. 2007). Oerter et al. (2007) describe a method for determining pyrite concentration by using a modified Rietveld method on heavy mineral separates. Clay minerals are best identified by XRD or EPMA (Moore and Reynolds 1989; Hall 2004).

Another approach that can be used to determine quantitative mineralogy is to calculate a normative mineralogy from the whole-rock chemical composition. A normative mineralogy is a set of idealized minerals that are calculated from a whole-rock chemical analysis; not all calculated minerals are always actually present in the sample (Neuendorf et al. 2005). Normative mineralogy is usually calculated for volcanic rocks that have glass and/or very small crystals, for which it is difficult to optically determine a modal mineralogy. A normative mineralogy is also calculated for altered or metamorphosed igneous rocks that no longer have the original igneous mineralogy. One of the most widely used normative calculations is the CIPW (Cross, Iddings, Pirsson, and Washington) norm. Another normative method involves the combination of XRD and whole-rock chemical analysis and attempts to provide better quantitative mineralogy (e.g., SedNorm; Cohen and Ward 1991; Ward et al. 1999). Combustion-infrared analysis for carbon and sulfur phases following various sample treatments is often useful in interpreting XRD patterns, due to greater sensitivity.

ModAn is a normative calculation that estimates modes "by applying Gaussian elimination and multiple linear regression techniques to simultaneous mass balance equations" (Paktunc 1998, 2001) and allows location-specific mineral compositions to be used. In a study at the Questa mine in New Mexico, representative mineral compositions for mineral samples were determined from EPMA and used in ModAn (McLemore et al. 2009b). A feature of ModAn that makes it appropriate for environmental studies is that it can model one of several sulfide minerals.

Automated SEM-based quantitative mineralogical analysis (e.g., MLA, QEMSCAN) is another method that quantifies mineralogical parameters such as modal abundance, mineral assemblages, texture (e.g., structural alignments), microfractures, microveins, and porosity in solid samples (Pirrie et al. 2004, 2009; Hoal et al. 2009a, 2009b). It is commonly used for metallurgical, mineral processing, and geometallurgical applications and shows promise for mineral-environmental applications. For example, it can be used as a complement to ABA and humidity cell testing. The method is based on an integrated system that includes a scanning electron microscope and proprietary software, and enables quantitative chemical analysis of materials and generation of high-resolution mineral or element maps and images (see Hagni 2008; Hoal et al. 2009b; Pirrie and Rollinson 2011; Smith et al. 2013; see Appendix 3 for additional information). This automated method produces statistically representative data sets due to the large number of high-speed determinations (approximately 150 per second).

Other techniques have been developed that can provide petrographic, chemical, and mineralogical information on ore, mine wastes, and other solid materials. For example, laser ablation inductively coupled plasma mass spectrometry (LA-ICP-MS) can provide spatial information about metal concentrations in a solid matrix (e.g., Diehl et al. 2007, 2010). Data from LA-ICP-MS line scans can also highlight correlations between elements. For example, overlapping peaks of cobalt and nickel with iron may indicate that these two metals are coprecipitated within the same iron-sulfide mineral. Another example is an approach developed by Jambor and Blowes (1998), wherein XRD data are compared with petrographic observations to make mineralogical interpretations. Parbhakar et al. (2009) also developed a mineralogical ARD index that takes into account diameter/size, degree of alteration, and morphology of the acid-generating phase, content of neutralizing phases, and spatial relationship between acid-generating and acid-neutralizing phases.

Regional mapping of mineralogy can be accomplished using remote-sensing techniques coupled with field mapping (e.g., Bove et al. 2007a, 2007b; Rockwell 2011, 2013; Yager et al. 2013). Remote-sensing techniques can be very useful in prioritizing ARD cleanup in large areas by identifying iron- and aluminum-rich minerals that have specific pH-stability ranges (e.g., Swayze et al. 2000; Hauff et al. 2006). Although ground-truthing is generally necessary to verify remote-sensing data, the ability of these methods to assess large areas can be quite effective and can result in significant cost savings when compared with the costs of on-the-ground sample collection and characterization at a regional scale. However, these remote-sensing techniques only work in exposed areas because vegetation masks the measurements. Appendix 3 presents summaries of several remote-sensing techniques. A variety of regional- and national-scale geospatial data are available at the USGS Mineral Resources On-Line Spatial Data Web site (see Section 3.13).

3.4.6 Solid Extraction Techniques

Solid-extraction techniques are at the interface between solid characterization and water quality. Some extraction techniques relate to regulatory requirements (e.g., synthetic precipitation leaching procedure [SPLP], toxicity characteristic leaching procedure [TCLP], and meteoric water mobility procedure [MWMP]; see Appendix 3). The field leach test (Hageman 2007; see Appendix 3) was developed by USGS as a fast, simple, and cost-effective leach test to screen and prioritize mining wastes for their water reactivity. A variety of other short-term leaching techniques have been used to assess mine-waste materials (e.g., see Al-Abed et al. 2006). Selective-extraction techniques have been used to provide insights into element partitioning within operationally defined fractions in earth materials. Extraction methods attempt to fractionate the different pools of trace metals by using sequential extraction steps with chemical reagents of various strengths, reaction times, temperature, and pH conditions (e.g., Tessier et al. 1979; see Appendix 3). Bioaccessibility tests, which are extractions using surrogate biological fluids, have been used to evaluate uptake of metals under biological conditions (e.g., Ruby et al. 1993; Drexler and Brattin 2007). Toxicity field test kits and simple laboratory screening tests can be used to assess potential metal toxicity (e.g., Choate et al. 2008). See USEPA (variously dated) and Appendix 3 for additional information.

3.4.7 Microbial Sampling*

The kinetics of nitrogen, sulfur, metal cycling, and pyrite oxidation are often controlled by microbial activity. Microbes live in water as free-floating planktonic organisms or are attached to mineral surfaces within biofilms—within microniche environments defined by pore-scale carbon, gas, and moisture conditions. In highly productive environments, such ARD settings, thick mats of biofilm can develop that host multiple species of archaea and bacteria within a biogeochemical ecosystem (Druschel et al. 2004). Such a system may have spatial and trophic structure, which should be considered in the sampling plan where relevant. Therefore, the required sampling

* This section was written by James J. Gusek and Lisa Bithell Kirk.

MICROBES IN MINE WATER AND WASTE—COLLECTING AND MANAGING LIVE SAMPLES

Lisa Bithell Kirk, Enviromin, Inc.

It's alive! Although they are not visible in hand specimens, diverse microbial communities live free floating in water and attached to mineral surfaces by biofilms, where they directly influence the sulfur, nitrogen, and metal geochemistry of mined materials. Enzymes produced by these microbes greatly accelerate biogeochemical reactions of interest to mine-waste management, both good and bad—especially sulfide oxidation and sulfate reduction, redox changes in solubility of many metals, nitrogen speciation, and carbonate precipitation. There is a growing comprehension of the influence of microbial biogeochemistry and the resulting biotechnological opportunities to harness these processes for environmental and metallurgical management. These matters point to a critical need for care in sampling to (1) avoid introducing bias by mismanaging microbial influence on biomineralization during storage and testing, and (2) obtain desired experimental control by correctly describing microbial influence where it is relevant. The demonstrated value of bioremediation applied in wetlands in remediation, use of biofilm-based reactors in water-treatment systems, and application of biologically based systems in hydrometallurgy and process engineering show a reverse trend in the application of biotechnology through the mine life cycle, and some believe that more comprehensive characterization of native microbial communities and integration of their functional ecology into the biogeochemical models of solute release and attenuation in mine-waste facilities may improve mine-waste management. To that end, microbiological considerations need to be addressed during sampling. Even if microbiology is not a focus of a mine-waste study, investigators should consider oven-drying the samples at low temperature to remove bias introduced by biological activity following collection or the potential for ongoing biomineralization processes during storage. Comparison of samples that have been treated differently (some live, some dead) is especially problematic and needs to be addressed carefully.

Samples collected for microbial analysis must be collected using sterile techniques, and the advice of a qualified microbiologist or biogeochemist should be sought to do this properly. Care must be taken to plan ahead to order sterile sampling supplies and prepare a detailed sampling, storage, shipping, and analysis plan. At times (e.g., when drilling), maintaining truly sterile conditions (e.g., with alcohol and an open flame) can be practically impossible, and sanitized conditions with suitable controls to evaluate the magnitude and extent of contamination must be an acceptable alternative. Where possible, drilling should be dry, with no use of additives involving carbon compounds. Collection of organisms through filtration is possible, although the density of planktonic organisms can be quite low and a large volume of water may need to be filtered to obtain the needed biomass for analysis. It can be more effective to collect suspended sediment and extract attached organisms from it as well. Sampling geologists or environmental scientists often require specialized training. Additional time is required for microbiological sampling to avoid contamination; this is especially true if collection of strict anaerobic organisms is desired because samples will need to be collected and preserved under nitrogen gas. Unique health and safety considerations arise from these sampling protocols, which should be considered in the industrial work environment. A good review of subsurface microbiological sampling issues is provided by Amy and Haldeman (1997).

Samples to be used for live cultivation (e.g., enumeration, rate reactors, enrichments, isolation experiments) should be collected using sterile protocol and stored under dark conditions at 4°C (39°F) until they can be shipped to the laboratory. Samples to be studied using molecular methods (e.g., analysis of genomic DNA, expression of enzymes using RNA) should be collected using sterile protocol appropriate to the desired analysis and flash frozen on-site using a dry ice and ethanol bath, and immediately transported on dry ice to a –80°C (–112°F) storage facility. Shipping can pose unusual challenges because freezing and thawing of nucleic acids is damaging; therefore, delays should be avoided wherever possible. International shipping of live samples for research and analysis purposes is also uniquely challenging and requires advanced planning.

methods will depend on the characteristics of the studied environment and on the dynamics of the targeted microbial community, as well as the goals for analysis of the samples (see Microbes sidebar).

As the kinetics of pyrite oxidation are mostly controlled by microbial activity, sampling of rock for bacterial characterization or quantification may be included in a sampling and analysis plan. Specific sampling techniques will be governed by the situation (e.g., is the original rock sample submerged in liquids or dry) and the sampling technique (hand sampling or drill cuttings). In drill-cutting-derived samples, the drilling mud composition can bias the sample either positively or negatively for acidophilic microbes. Positive bias could be introduced through recirculating mud or cuttings that are developed nearer the surface of a given hole; negative bias may be introduced by using organic-based muds containing polymers that could support bacterial communities that can out-compete *Acidithiobacillus ferrooxidans*. Extraordinary care may be required in sterilizing drill bits and minimizing drill-cutting recirculation to obtain unbiased rock samples for microbial characterization. Microbiological sample preservation techniques are beyond the scope of this discussion, and it is assumed that a qualified microbiologist would be responsible for developing the sampling and analysis plan.

3.4.8 Sampling Theory and Errors

The final results of an analytical determination will never be any more reliable than the reliability of the sampling. In fact, sampling is the most important step in an analysis. Many modern analytical techniques have errors <5%, whereas poor sampling can result in errors ≥100%. If a sample was not properly collected or is not representative of the target population, then it doesn't matter how much QA/QC is performed by the laboratory or during other aspects of the investigation.

All materials to be sampled exhibit some degree of heterogeneity, which is important because overlooking, or unfamiliarity with, this heterogeneity can result in significant sampling errors, incorrect decisions, and wasted effort and expense. Sampling theory is applied as a preventive tool to find ways to minimize the negative effects of heterogeneity on sampling. The degree of heterogeneity depends on the scale of observation. This concept is illustrated by the pile of sand shown in Figure 3.5. From a distance, the sand appears to be homogeneous, but closer examination reveals the heterogeneity of the particles. Scale (both spatial and temporal) is a very important consideration. Variability at one scale may be negligible at a different scale; hence, the scale of sampling for an investigation must match the scale of the objectives of the investigation and of the ultimate application of the data (Smith 2007; see Understanding Scales sidebar).

Field sampling, sample preparation, and laboratory subsampling are generally the major sources of error in environmental investigations (Gy 1998; Ramsey and Hewitt 2005). Therefore, DQOs need to take into account sampling, sample preparation, and analytical errors when defining the sampling plan, scale, and desired uncertainty. The theories of Pierre Gy (1982, 1998) present practical sampling, comminution, and subsampling methods that can be applied for less expense because fewer samples are necessary with this sampling approach. Careful attention to these techniques of sampling correctness can result in samples that better represent the site and data that more truly represent the sample.

Sampling theory addresses two types of heterogeneity, compositional (constitutional) and distributional. Compositional heterogeneity refers to the variability in the composition of individual components of the material. For solid materials, it cannot be remedied by mixing; it can only be controlled by increasing the sample mass or physically changing the material (i.e., decreasing the particle size). The compositional heterogeneity error is greatest when the analyte is present as a few discrete particles of pure material (e.g., discrete gold particles, sometimes called the *nugget*

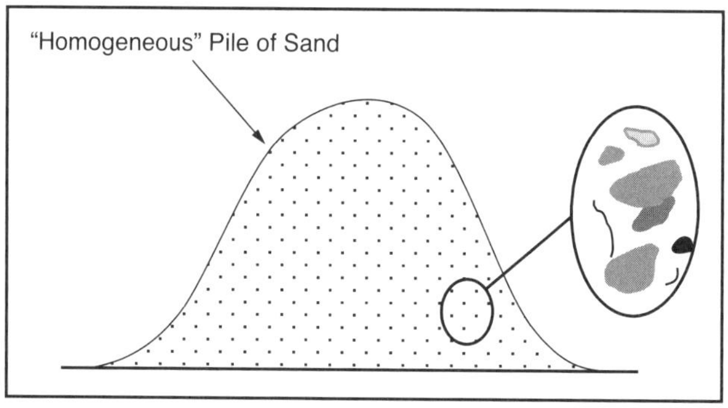

Source: Adapted from Myers 1997.

FIGURE 3.5 Heterogeneity as a function of the scale of observation

effect in exploration). Distributional heterogeneity is the spatial component of the variability. It is related to the nonrandom spatial and/or temporal distribution of the components of the population.

Gy's sampling theory (Gy 1982, 1998; Pitard 1993; USEPA 1999) addresses seven types of errors:

1. **Fundamental sampling error (FSE):** This error is due to compositional heterogeneity in the chemical and physical composition of the material (includes particle size distribution). It can be reduced by decreasing the diameter of the largest particles (comminution) or by increasing the sample mass. In the field, increasing the sample mass is usually the only practical option (unless drilling results in particle size reduction).
2. **Grouping and segregation error (GSE):** A GSE is due to nonrandom distribution of particles (i.e., distributional heterogeneity). This error can be reduced by collecting and combining many randomly selected increments.
3. **Long-range heterogeneity error (shifts and trends):** This is a fluctuating, nonrandom spatial error.
4. **Periodic heterogeneity error (cycles; QE2):** This is a fluctuating temporal or spatial error.
5. **Increment delimitation error (IDE):** An IDE is due to the wrong choice of sampling equipment. It can be reduced by using well designed sampling equipment.
6. **Increment extraction error (IEE):** This error is due to the inappropriate use of sampling equipment. It can be reduced by using well-designed sampling equipment and good sampling protocols.
7. **Preparation error (PE):** This error is due to loss, contamination, or alteration of a sample. It can be reduced by using appropriate sample handling, processing, and preservation, and by choosing appropriate analytical methods.

When a sample is collected, the amount of the FSE is due to the compositional heterogeneity (e.g., mineral composition, shape, liberation, and particle size) of the material being sampled. The FSE never can be eliminated, but it can be minimized and estimated. Procedures to minimize the FSE need to be included in the DQO process and the FSE should be estimated prior to sampling. It is important to keep the variance of the FSE within reasonable limits. The variance of the FSE can be estimated using the following formulas:

UNDERSTANDING SCALES OF GEOLOGIC SYSTEMS

Richard B. Wanty, U.S. Geological Survey

In addition to using proper sampling protocols, it is important to have a comprehensive sampling strategy designed to meet the needs and goals of the project. Such strategies require a considerable amount of thought prior to collection of the first sample. Strategies are tempered by practical considerations such as budgets (of time and money), physical access to sampling locations, and availability of equipment, but they should be driven primarily by the question: What do I hope to learn about the study area?

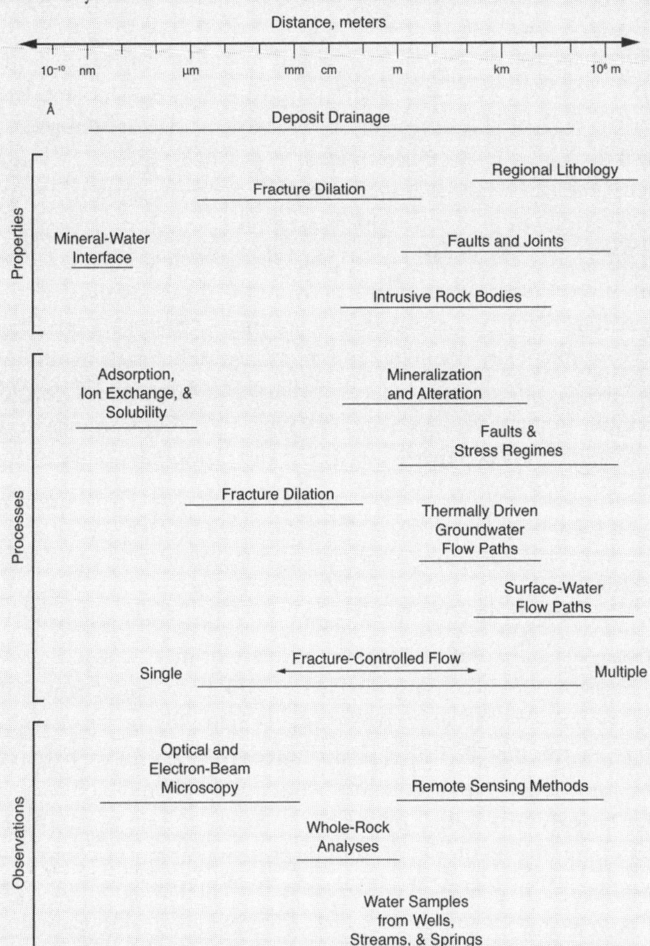

A subsidiary question might be: If I collect a sample {of something}, what does it represent? To answer this question adequately, one must consider the spatial and temporal scales at which geologic and chemical systems vary, and plan sample location and timing accordingly.

In the adjacent figure, the number line across the top represents 16 orders of magnitude of distance, from 10^{-10} to 10^{6} m. A variety of properties and processes of geological systems are compiled, as well as some common techniques used in field campaigns, along with horizontal lines to represent the approximate limits of the spatial scales relevant to each item. By lining up the properties and processes that are relevant to the study with the observational methods in the bottom portion of the figure, a more efficient sampling strategy can be developed. In many cases, each observational method represents a particular specialist, so application of this figure can also aid with project staffing.

The figure could easily be modified to include other properties, processes, and observational methods, but the example given here is simply designed to illustrate the approach. Additionally, one might easily envision a similar figure in which the number line across the top represents temporal, rather than spatial, scale.

$$\sigma^2_{FSE} = (K\, d^3) / M_S$$

$$\log \sigma^2_{FSE} = \log K + 3 \log d - \log M_S$$

where K is an empirical constant, d is the maximum particle diameter (in centimeters), and M_s is the mass of the sample (in grams). As shown in the previous equation, the value of log σ^2_{FSE} (the variance of the FSE) is proportional to –log M_S. Pitard (1993) provides nomographs that relate the variance of the FSE to the maximum particle diameter and the sample mass. These nomographs can be used to optimize the sampling protocol.

From the time of collection to the time of analysis, a sample may undergo preparation involving several splitting or subsampling events, and each of these events gives rise to sampling error. For solids, the FSE can be minimized by increasing the sample mass or decreasing the particle size of the largest particles. Also, it is critical to use proper sampling equipment and protocols, which addresses the IDE, IEE, and PE. Once the correct sampling equipment and protocols are determined, it is possible to minimize the GSE by collecting multiple increments (i.e., subsamples). In general, a sample should be composed of at least 30 increments. In practice, especially for trace constituents, 50 to 100 increments should be used. Random sampling is also important, and it is essential that all of the target population be equally available for sampling.

Pitard (1993) and Myers (1997) provide several examples of the advantages and disadvantages of different designs of sampling equipment. Unfortunately, most of the sampling devices currently available for environmental sampling are incorrect (e.g., augers, thiefs; Myers 1997). Pitard (1993) points out that pipe samplers, such as augers and thiefs, are problematic because the bottom portion of the sample is under-represented due to the presence of the conical head. The U.S. Army Corps of Engineers Cold Regions Research and Engineering Laboratory has developed a Multi Increment sampling tool (CMIST) that is consistent with sampling theory (Walsh 2009). In addition, Pierre Gy recommends scoops and spatulas that are flat with parallel sides to avoid preferential sampling of coarse particles (Gy 1982, 1998; see Appendix 3).

3.4.8.1 Multi Increment Sampling

Multi Increment (MI) sampling (EnviroStat, n.d.) is a composite sampling approach based on sampling theory that reduces data variability and provides an estimate of mean concentrations within a target population (or decision unit; Ramsey and Hewitt 2005). An increment is the material extracted from the target population in a single operation of a sampling device. A sample is obtained by the combination of several increments, with the objective to minimize heterogeneity and to reduce the GSE. To properly represent the entire target population, material needs to be sampled from many locations within the target population (Wait and Ramsey 2007). MI sampling is similar to incremental sampling methodology and incremental-composite sampling, and involves collecting numerous (typically 30 to 100) increments of soil that are combined, processed, and subsampled according to specific protocols (see ITRC 2012; see also training resources listed in Section 3.13). MI sampling reduces analytical costs compared with collecting and analyzing multiple discrete samples. The State of Hawaii Office of Hazard Evaluation and Emergency Response discusses MI sampling in its online technical guidance manual (Hawaii Department of Health 2009). The State of Alaska has prepared draft guidance on MI sampling (Alaska 2009). In their guidance they point out that "MI may not be used to dilute contamination and therefore underestimate the need for cleanup. This may occur if the decision unit inappropriately incorporates large, uncontaminated areas in addition to real source areas" (Alaska 2009).

A potential disadvantage of MI sampling is that information about spatial variability within the target population is lost because the target population is represented by one MI sample.

However, collection of triplicate MI samples allows for statistical calculation of the standard deviation, the relative standard deviation, and the 95% UCL (upper confidence limit) of the mean concentration for a given target population (or decision unit). Therefore, variability in spatial concentrations between different target populations can be evaluated, which requires careful definition of the target populations (or decision units) at the beginning of the investigation. The 95% UCL of the mean can be an important statistic because it is often used in risk assessments.

3.4.9 Sampling Strategies and Sampling Designs

During the seven-step DQO process (USEPA 2006b), it is necessary to define the goal of the investigation, which states how environmental data will be used in meeting objectives and solving the problem, identifies study questions, and defines alternative outcomes. This information dictates how the sampling strategy will be handled. For example, which of the following questions is more applicable:

- Is a particular contaminant present at the site? or
- What is the average concentration of a contaminant in a particular area?

These are very different questions that necessitate very different approaches to sampling.

Figure 3.6 illustrates reproducibility and bias in sampling where there is one sample collected for a given location or time. The top two graphs illustrate situations in which the mean sampling error and the sampling variance are near zero, which is the goal of proper sampling. The bottom two graphs illustrate problem situations in which the sampling is not reproducible. These problem sampling situations can lead to costly mistakes and flawed decisions.

Correct sampling is the only way to eliminate sampling biases (Gy 1998). For correct sampling, all of the components of interest in the target population must be available and must have an equal probability of being randomly selected for inclusion in the sample. Therefore, both randomness and equiprobability must be considered when designing spatial sampling. Myers (1997) lists the following key components of a sampling strategy:

- Select the target population.
- Characterize the heterogeneities.
- Develop a sampling plan.
- Implement the sampling plan.
- Characterize spatial variability.
- Optimize within DQO constraints.

Cochran (1977), Gilbert (1987), and Gilbert and Pulsipher (2005) discuss the details of several sampling designs and their application for different sampling objectives. Sampling designs have different strengths and weaknesses; consequently, the chosen sampling design should be consistent with the site, media, and presence of trends or cycles. Commonly used sampling designs include strict random sampling, systematic random sampling, and stratified random sampling. There also are sampling methods that don't involve random selection, including convenience sampling, judgmental sampling, and purposive sampling.

In systematic random sampling, samples are evenly distributed over the entire target population in an organized manner. Systematic grids can include square, rectangular, or radial grids (Figure 3.7). Systematic random sampling has a random start with subsequent samples taken at a set interval. If any periodicities or cycles exist, the sampling interval needs to be evaluated so that it does not result in biased results. Systematic random sampling has the lowest error of all random sampling protocols and is the easiest to implement.

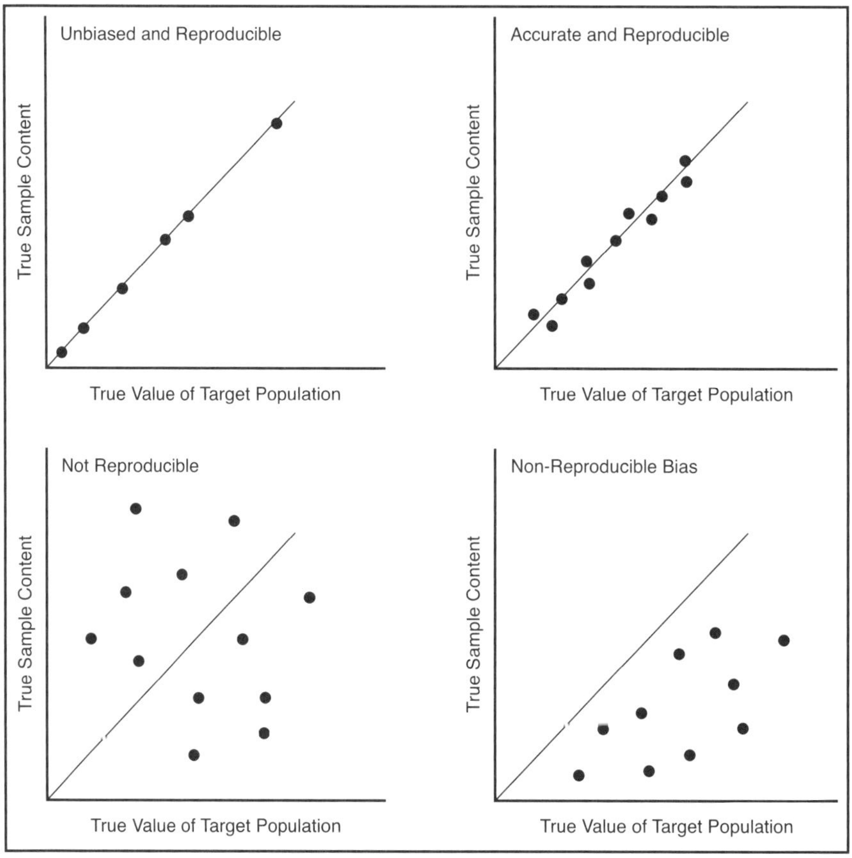

Source: Adapted from Pitard 1993.

FIGURE 3.6 Reproducibility and bias in sampling

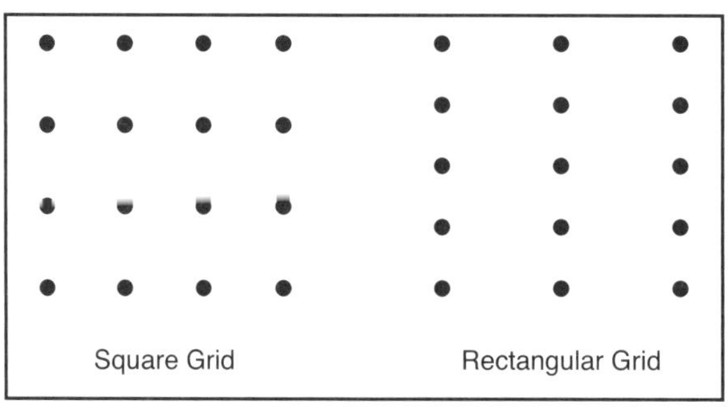

Source: Adapted from Myers 1997.

FIGURE 3.7 Different types of grids used in systematic random sampling

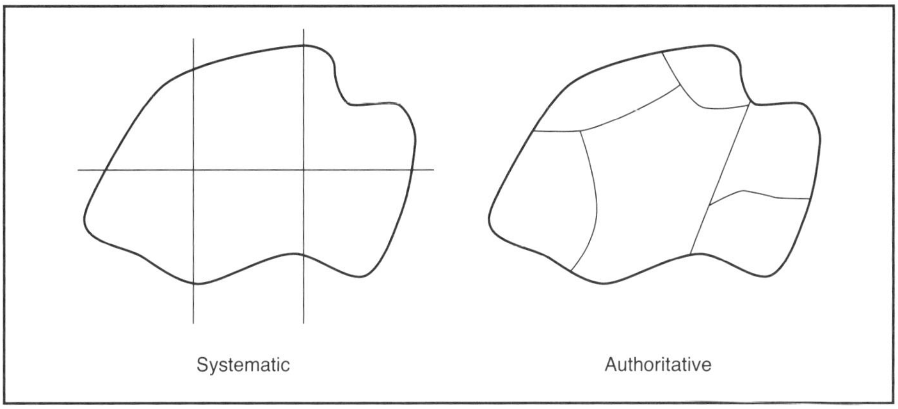

Source: Adapted from Myers 1997.

FIGURE 3.8 Types of stratified random sampling

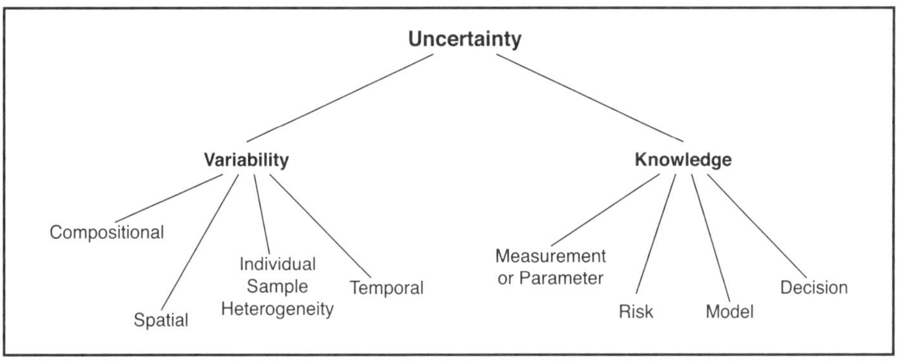

Source: Adapted from Haimes 1998.

FIGURE 3.9 Some sources of uncertainty

Stratified random sampling involves some grouping of the sample population and then randomly sampling within each group (Figure 3.8). It has both components of randomness and equiprobability. Stratified random sampling is often used when the target population can be split into non-overlapping subgroups that individually are less heterogeneous than the target population as a whole (but this is not a requirement). If there are no obvious subgroups, grid lines can be used to divide the target population (Figure 3.8). Random samples are taken from each subgroup and the results are combined to estimate parameters of the population. However, this approach does not lend itself to MI sampling because there is not enough mass or increments in each subgroup to represent it.

Figure 3.9 summarizes several sources of uncertainty. The desired degree of certainty for an investigation must be decided at the inception of the investigation. A low degree of certainty in the data can lead to erroneous interpretations and flawed decisions. Conventional thought is that a higher degree of certainty generally requires more sampling and higher costs. However, better sampling techniques that incorporate sampling theory can provide more representative data at reduced costs.

3.5 DRILLHOLE/BOREHOLE SAMPLING CONSIDERATIONS*

3.5.1 A Reasonable Sampling Plan

A senior exploration geologist would typically develop a drillhole sampling plan, whether it is for exploration borings or production blastholes. The plan needs to consider drilling production because drillers may be compensated on depth penetrated. If drill bit advance is interrupted by sampling operations, then standby charges, which can be onerous, are certain to be incurred. Worse, an unhappy driller can compromise sample integrity in a number of ways that might only be discovered weeks after the fact, if at all.

Once the sampling plan is understood and accepted by the drill crews, a qualified geologist or engineer will periodically oversee the drilling program to examine borehole cuttings and disturbed material, and should be on call to resolve issues that can arise. Field inspection of materials is recommended for all components of the project, though especially for those components of the project in medium- to high-ARD risk zones or where potentially acid generating (PAG) rock or acid-producing rock (APR) materials have been identified. The geologist/engineer may examine the materials for

- Mineralogical content, in particular the presence of sulfides or siderite and other carbonates;
- Fizz rating;
- Sulfur odor, particularly during fizz testing;
- Paste pH; and
- Steam from completed drillholes.

If low paste-pH values (e.g., below 5.0; Skousen et al. 1987), a lack of fizz (Sobek et al. 1978), sulfurous odors, steam, and sulfide minerals are observed, then materials can be assumed to be PAG or APR and should be tested and/or managed appropriately. If some, but not all, of the signs of ARD are present, they should be noted in the drill logs and precautions taken to prevent ARD formation. The geologist or engineer overseeing the project may need to inform site personnel (e.g., the chief project/site geologist or geochemist) of these tests and their proper implementation to allow appropriate classification and handling of materials.

3.5.2 Borehole Sampling Considerations

If PAG/APR zones are identified during the preliminary exploration drilling program, then blast borehole sampling should be performed for ARD potential to ensure that these materials are accounted for and managed appropriately during active mining of the deposit. Until a less-prescriptive blasthole sampling program that can be understood and followed by drillers is developed by an experienced site geologist/geochemist, the following guidance is offered as a starting point. A minimum of three boreholes per blast should be sampled for environmental purposes, or as many as are necessary to obtain optimum coverage of the material to be blasted. Samples should be collected a minimum of every 1.5 to 2 m (5 to 6.6 ft), or a certain percentage of the samples collected for routine ore and waste control, if applicable. However, if significant lithologic changes are observed, samples may need to be collected at smaller intervals.

Samples should be immediately submitted for static ABA and paste pH analyses (e.g., 48-hour turnaround times or less) because once the holes are loaded with blasting agents and detonated, excavation to either the waste rock dump or ore stockpile(s) can shortly follow. If acid-producing mine production waste is to be properly managed, especially in zones that may or

* This section was written by James J. Gusek.

may not be acid producing, the destination of these materials needs to be known well in advance to avoid disruptions in haulage routes and fleet scheduling. Once a truck load of APR is dumped in a *clean* rock dump instead of in a planned APR repository, it is very difficult to undo because of contamination of the non-acid-producing material with the AD from the APR.

3.5.3 Sampling Methods

This section provides guidelines for appropriate sampling methods for rock and water that may be intercepted during drilling activities, which is not to be confused with water sampling from properly completed and developed monitoring wells. The purpose of sampling is to obtain representative samples of the material throughout the entire area that will be affected by project activities. The collection of representative samples is the primary intent of the sampling methods described in this section. Sample handling, logging, shipping, storage, and chain of custody are also described. Sampling should be performed by or under the oversight of a qualified scientist or engineer.

3.5.3.1 Rock Sampling

Rock and soil samples may be collected into resealable plastic bags or other clean, sealable sample receptacles (e.g., bottles or ore bags). Samples should be logged using standard site procedures or methods. Generally, a 4-L (1-gal) resealable plastic bag filled with material is sufficient volume for recommended laboratory tests; however, this volume may be adjusted by the analytical laboratory performing the work or the geologist/engineer requesting analyses. Microporous vapor-permeable bags can be useful for collection of development drilling or blasthole samples to permit drying of the samples in the bags without loss of fine solids. Drill samples containing particulate gold may require significantly larger samples (up to 15 kg [33 lb]) if gold assay is one of the sampling objectives.

Once collected, samples should be sealed to prevent oxidation or moisture loss if extent of oxidation and moisture content is to be determined. Typically, sample containers are labeled with information such as

- Date and time of collection,
- Borehole identification,
- Depth interval sampled,
- Geologic formation (if known or suspected),
- Alteration (if known),
- Individual conducting the sampling,
- Station number,
- Project identification, and
- Other pertinent information.

An appropriate number of duplicate samples should be collected to meet the project DQOs. Samples should be analyzed at a certified laboratory if the results are needed to meet permit or compliance requirements and always analyzed using appropriate QA/QC measures. Sample locations should be mapped and given a unique identifier. Locations and sample identification should be logged into field notes and entered into the geographic information system database, if applicable.

3.5.3.1.1 Drilling Methods

Drilling for geochemistry samples for APR evaluation may be combined with geotechnical drilling. The following drilling methods are examples of acceptable alternatives for collection of rock or soil samples over the entire mine life. Whereas some methods are preferred because they provide more distinct and intact samples, this must be balanced with time and expense. The following methods are listed in order of preference for collection of geochemistry samples for APR evaluation:

- Conventional coring
- Roto-sonic coring
- Air rotary reverse circulation
- Air rotary

One drawback of the air rotary methods is that salting of the hole can occur. As the drill penetrates a rock layer, the rock chips are blown out of the hole by the air drilling process. In some cases, the pyrite-containing rock chips are heavier and tend not to be adequately exhumed from the drillhole. Subsequently, these chips fall back into the hole, biasing the amount of pyrite in the next sample. This can usually be detected by reviewing the test results. If the tests indicate an increasing amount of APR as the depth increases, then re-drilling and testing should be considered, or a review of the test results and rock samples may be required. A drawback of core drilling is that fines can be washed away before the core is retrieved from downhole, which can introduce significant bias into subsequent analytical procedures if critical minerals, such as sulfides or other trace metals, are associated preferentially with finer material.

Auger sampling is generally not recommended for geochemical sampling of rocks, as intermixing and contamination of samples can occur as materials are lifted by the auger screw (Kleinmann 2000). However, auger sampling of soils from shallow depths (i.e., less than 3 m [about 10 ft]) may be acceptable if soil stratigraphy or depth to bedrock is well understood. Samples can be collected from drill cuttings or the drill core. If difficulties in determining sample depth or the geologic strata are encountered because of incomplete or inefficient cuttings return or cuttings are significantly altered, then coring may be required. Samples should be collected dry when possible. In this handbook, it is assumed that the drilling fluids are derived from uncontaminated water sources; if the water source is suspect, then it should be sampled along with the rock core.

3.5.3.2 Water Sampling

Exploration drilling and mining operations can encounter groundwater. Drillers may note zones of excessive groundwater inflow in the drilling mud management or appearance of moisture changes in drill cuttings. It is virtually impossible to collect an unbiased sample of these waters, but water chemistry might be inferred by spot sampling the drilling mud, filtering out the suspended solids, and comparing the chemistry with a known baseline.

When drilling with air, drillers can pause at a zone that is *making water*, stop drill bit advance, and blow water out of the hole. Sampling and analyzing this water may provide clues to the chemistry of the rock mass being penetrated, and the results may suggest zones for future monitoring wells.

3.5.4 Sample Storage and Chain of Custody

Following logging and collection, any rock samples collected primarily for environmental purposes should be stored in a closed, water-resistant, noncontaminating receptacle (such as a cooler) and kept in accordance with the site-specific sampling plan protocol until transfer to the analytical laboratory. Routine ore and waste-control samples collected primarily for metallurgical purposes should be handled in a manner that optimizes assay turnaround requirements. Unlike water samples, which are cooled in transit to analysis laboratories, rock sample temperature will probably be governed by the ambient temperature of the ground mass, and use of an insulated cooler will likely preserve that condition at least temporarily. Rock samples should be shipped or transported to the analytical laboratory as necessary to ensure that holding times, if any that may be internally imposed, are not exceeded. For example, it might be poor policy to accumulate rock chip samples from a given blast round for ABA parameter analysis past the planned date/time that the round would be loaded and shot, and excavation would commence shortly thereafter. The holding times should consider typical laboratory turnaround times in order to receive the analytical data in time to make reasonable characterizations.

Samples should be stored and shipped following site-specific chain-of-custody procedures using guidance from ASTM D 4840-99 or the analytical laboratory. Doing this would help to ensure that the intended test methods and protocols are applied to the samples submitted. The chain-of-custody forms also provide a helpful paper trail for sample tracking.

3.6 BIOLOGICAL SAMPLING CONSIDERATIONS—AQUATIC*

Direct measurement of the biological condition of the aquatic environment in areas influenced by mining is a critical component in understanding the potential and realized ecological impacts of that human activity. If correctly employed, biological monitoring can help to answer many key questions about mining impacts, such as the following:

- What organisms would we expect to be present at a given site absent mining or processing?
- Are organisms that are expected to be present currently resident at a given site year-round or seasonally?
- How healthy is the aquatic ecosystem at a given site?
- Is poor water or sediment quality resulting from ARD a direct limiting factor for the aquatic ecosystem?
- What other factors (e.g., lack of suitable habitat) could be limiting factors for the aquatic ecosystem at a given site?

Collection of biological media is discussed in Appendix 3.

Aquatic communities present at a given location are a function of the biological, physical, and chemical conditions that exist at that site, all of which may vary over time. In order to answer questions about the potential impacts of mining and processing on surrounding aquatic resources, it is important to design studies in an effective manner. Several experimental approaches exist to evaluate an aquatic ecosystem's response to a stressor such as ARD (Power 2002). For example, to gauge the impact of a specific mine or process-related discharge on the receiving stream, comparison of the downstream aquatic community with an unimpacted upstream reach or reference site(s) may be appropriate. In order to evaluate the downstream extent of a mining impact, measurement of the biological community response to decreasing exposure severity with increasing distances downstream could be completed. In a temporal context, sampling the aquatic community before and after a mine remediation project could serve to document the benefits of that project. Where possible, identification and utilization of an analog watershed (i.e., a watershed that is lithologically similar, absent the mining disturbance) can be useful in determining the

* This section was written by Andrew S. Todd.

premining geochemical and ecological condition. Importantly, the characterization of a mine's impact or the development of a mine remediation goal relying solely on biological population/community data is not advised, and the collection of additional abiotic analytical data (see other sections of this handbook) is highly recommended to support these population/community data.

A typical first step in evaluating the *health* of an aquatic ecosystem is the characterization of its expected biological condition, absent any direct anthropogenic influence (e.g., mining). Typically, the expected biological condition is established through the quantification of biological community metrics within an appropriate *reference* site(s), which can be an unimpacted upstream reach, or from a site in a nearby watershed with similar physical characteristics (including similar lithology). Comparisons of resultant biological information obtained from the presumed unimpaired reference site with data from the test site can be used to assess degree of impairment. The establishment of good reference sites is paramount in this approach to ensure that sites would have comparable biological potential absent the anthropogenic stressor of interest (Stoddard et al. 2006). For example, a lack of fish at a well-selected reference site might suggest that a measured lack of fish at the test site is a consequence of poor habitat rather than poor water quality (Milhous 2007). As such, regardless of the apparent suitability of a reference site, some level of habitat characterization is advised in both reference and impacted reaches to ensure reliable interpretation of the results of their comparison. Habitat comparisons can range from qualitative to highly quantitative depending on one's needs.

Several biological assemblages (e.g., algae, benthic macroinvertebrates, fish) are suitable for characterizing ecological stress in aquatic systems. Overall study design and other practical considerations (e.g., budget, schedule, access) will ultimately define what biological communities can/should be sampled in a given water body. In general, employment of multiple lines of biological evidence is recommended to thoroughly characterize the nature and extent of impairment and/or risk (Besser et al. 2007; Farag et al. 2003). For example, in an ARD-impacted stream devoid of fish, electrofishing alone would not necessarily yield a likely cause of impairment; instead it would simply identify the impairment itself (i.e., lack of fish). However, in conjunction with analyses of exposure media for contaminants or with site-specific water and sediment toxicity testing conducted within a tightly controlled laboratory environment, it would be possible to establish if poor water or sediment quality (as opposed to other potential causes, such as lack of available food or habitat) was the cause of the observed impairment. It is important to independently evaluate all potential causes for the observed biological condition, as impairment could be caused by any (or all) of these potential limiting factors.

Directly sampling fish populations can yield a variety of useful information about the causes and relative degree of biological impairment. Qualitative fish sampling, through methods such as hook-and-line fishing or *spot* electrofishing, can be a cost-effective means of evaluating the presence or absence of fish in MIW. Quantitative fish sampling methods, such as multiple-pass depletion electrofishing protocols, produce fish population metrics (e.g., population size, species composition, biomass), which can be compared with reference-stream measurements (Farag et al. 2003; Reynolds 1996; Bervoets et al. 2005). Measurements taken on individual organisms can also be used as a line of evidence in documenting an impaired aquatic ecosystem. For example, simple measurements of fish lengths and weights can be used to calculate and then compare average condition factors for groups of individual fish (i.e., fish this long should weigh this much) from impacted and reference waterways (Todd et al. 2007). Further, fish captured during surveys can be employed through the use of standard fisheries techniques (e.g., tagging, radio telemetry) to answer additional important questions about the fish population (i.e., is the fish population resident year-round?) (Hoar 2005; Goldstein et al. 1999; Scullion and Edwards 1980). Finally, stocked fish can be utilized in the field to evaluate the site-specific suitability of water quality conditions through the use of in situ exposure experiments (e.g., Todd et al. 2007; Farag et al. 2003; Harper et al. 2008).

In the laboratory, fish can be used to develop additional information in isolating the root cause of observed impairments. For example, toxicity testing is often conducted to establish threshold levels of a given contaminant of concern (e.g., Zn) to a given organism (e.g., rainbow trout), with these thresholds then compared to field measurements of the contaminant to assess risk to the organism from that contaminant (USEPA 2002b). While this type of traditional reference toxicity testing can prove useful in many ways, because of the chemical complexity of most site waters and resultant implications for toxicity, toxicity testing using serial dilutions of site waters can also yield relevant, site-specific toxicity information (Besser and Leiby 2007; USEPA 2002b). Fish tissues have also been analyzed in the laboratory to serve as lines of evidence linking metal exposure pathways to the physiological malfunction in the organism. For example, tissue metal concentrations have been employed to serve as a link between elevated aqueous metals concentrations and observed fish impairment (Todd et al. 2007; Farag et al. 2003; Bervoets et al. 2005). This tissue-residue approach does not always yield useful information, as metals concentrations can be so acutely toxic that tissue concentrations do not have a chance to build before the organism perishes. Additional biochemical measurements can be used as measures of metal exposure, including the measurement of lipid peroxidation (e.g., Farag et al. 2003) and the metal-detoxification protein metallothionein (e.g., Hamilton et al. 1987).

Measurement of lower aquatic trophic levels (e.g., benthic macroinvertebrates, algae) can also be an important component to studies evaluating the ecological impacts of past, present, or future mining in a given area (e.g., Griffith et al. 2004). Largely because of their widespread presence in aquatic environments, their limited mobility, and their ease of collection, benthic macroinvertebrates themselves are frequently employed as direct indicators of aquatic ecosystem health as community parameters such as organism diversity and abundance respond to anthropogenic stressors such as ARD. Specifically, benthic macroinvertebrate metric categories have been defined (e.g., taxa richness, species composition, pollution tolerance, functional feeding groups) to represent assessments of community structure and function, and appropriate combinations of these metrics can yield a useful index characterizing the health of the resident aquatic community (Barbour et al. 1999). Although there are hundreds of metrics that can be used to assess community structure and/or function, a few (e.g., richness, abundance of mayflies, stoneflies, and caddisflies) have been found to predictably respond to increasing metal concentrations (Carlisle and Clements 1999; Clements et al. 2000). Benthic macroinvertebrate communities are routinely used as indicators of water quality and ecosystem health in both state and federal protocols, and many states are moving toward the adoption of benthic macroinvertebrate-based biocriteria for the assessment of attainment of the aquatic life use (e.g., CWQCC 2010). Similarly, algal community metrics (i.e., diversity, biomass) can be utilized to evaluate the ecological implications of a stressor such as ARD (Niyogi et al. 2002; Barbour et al. 1999).

A number of factors can strongly influence the character of a benthic invertebrate community observed between sample locations, even in the presence of a strong stressor (Clements and Kiffney 1995). For example, resident taxa are influenced by changes in elevation (Clements and Kiffney 1995), time of year (Clark and Clements 2006), and biogeography (Hawkins et al. 2000). As a result, it can be difficult to make accurate predictions regarding the character of a benthic macroinvertebrate community among sites sampled within a large study area (such as a state or ecoregion). Predictive models, which often account for differences in site characteristics that cannot be directly anthropogenically influenced (e.g., elevation, latitude, longitude, weather), can be used to assess differences in taxonomic completeness of benthic macroinvertebrate, fish, or algae communities, across these gradients (Hawkins et al. 2000, Carlisle et al. 2008). These models evaluate the relationship between the taxa that are observed (O) at a site with that which would be expected (E) to be present at that site (O/E ratios). In this manner, this approach can be used to attribute differences in O/E ratios to a change in the habitat quality of the sampling location. Increasingly, state and federal agencies are moving toward the use of such predictive models to measure biological impairment (Stribling et al. 2008; USEPA 2006c).

In the evaluation of ecological impacts from mining, lower trophic levels (i.e., algae, benthic macroinvertebrates) are useful beyond direct measurement of their community metrics. These organisms are often utilized in a similar manner as fish, including through use in lab and in situ toxicity exposures (e.g., Clark and Clements 2006) and evaluation of trace-metal bioaccumulation (e.g., Rainbow 2007; Winterbourn et al. 2000). Importantly, benthic macroinvertebrates serve as an important link in both stream and riparian food webs. As such, potential dietary toxicity pathways may be an essential component in evaluating mining- and process-related risk to higher trophic levels such as fish (e.g., Besser et al. 2001) or riparian birds (Custer et al. 2009).

Several important considerations must be recognized when considering the use of lower trophic levels such as benthic macroinvertebrates and algae. Primarily, both algal and macroinvertebrate communities have distinct seasonal cycles, and as such, care must be taken to ensure that sample collection dates are well documented to avoid inappropriate comparisons. Many sampling protocols recommend a single *index period* to avoid potential complications (e.g., baseflow for algae); however, this specificity will not always be appropriate, depending on the study design (Barbour et al. 1999). While fish populations typically demonstrate less seasonality, with exceptions including migration and/or spawning times, considerations for timing fish sampling might include safety (i.e., not during peak run-off) and targeted sampling during periods of the year demonstrating highest risk (e.g., highest aqueous metal concentrations). Additionally, in sampling aquatic biota, it is critical to recognize and plan for the diversity of aquatic habitat and resultant differences in expected aquatic assemblages. For example, the benthic macroinvertebrate community sampled from a pool is likely to be very different from that sampled from a neighboring riffle, so sampling representative habitat is important for obtaining meaningful results (Barbour et al. 1999).

Measurement of components of the aquatic biological community can be used as evidence in identifying ecological impacts from anthropogenic stressors, such as ARD. In measuring and interpreting the data from these biological communities, it is important to be mindful that these assemblages are a function of physical, chemical, and biological interactions integrated over time. As such, a given biological sample should be viewed as a snapshot of that community's condition at that given time, and multiple lines of evidence should be developed to support any conclusions regarding the causes of observed impairment.

3.7 AIR SAMPLING CONSIDERATIONS*

There are a variety of reasons to sample air quality at mine sites including monitoring for combustible gases or vapors (e.g., methane), hazardous gases, vapors, aerosols (e.g., hydrogen cyanide, mercury, sulfur dioxide, and acid from solvent extraction–electrowinning [SX-EW] plants), particles (e.g., mercury salts, silica dust, diesel, asbestos, dust), oxygen deficiency, total suspended particulate matter or particulate matter <10 µm (PM_{10}) levels in ambient air, individual exposure

* This section was written by Karl L. Ford and Susan J. Caplan.

AIR EMISSIONS

Virginia T. McLemore, New Mexico Bureau of Geology and Mineral Resources

Airborne emissions occur over the entire mine life cycle and ultimately can affect air and water quality, personnel health, health of fauna and flora in the environment, and can cause increased wear to machinery (AGDR 2009). During activities such as blasting, extraction, and processing, dust and larger size particles derived from the mechanical breakdown of rocks and soils are dispersed by the wind. These particles can oxidize or alter during transport before being deposited, in many cases within local water supplies. Gas emissions from fuel combustion or mineral processing can include sulfur dioxide (SO_2) and nitrogen dioxide (NO_2) and ultimately affect water quality. Mercury can be released in some mines and processes, especially in some gold operations, including autoclaving.

Although most mining and processing sites in the United States will require extensive sampling and monitoring of air quality in order to meet national and local regulatory requirements for permit application and the National Environmental Policy Act, additional sampling and monitoring could be required in order to evaluate whether air quality could adversely affect water quality, such as contributing to acid drainage. Local meteorological stations and data will also be necessary for air resources analyses to enable computer modeling of potential air-quality impacts (see Appendix 3). The station should record wind speed, wind direction, rainfall, temperature, evaporation, barometric pressure, and humidity. During mining, beneficiation, and mineral processing of most commodities, particulate material and gases can be emitted that can affect water quality and soils downwind of the mining facilities. Particulate materials include flue dusts (from sinter, roaster, smelter, or refinery stacks) and fugitive dusts (from crushers, tailings facilities, road use, blasting). Most of these are regulated in current mining operations by the U.S. Environmental Protection Agency and state and local government agencies; however, in the past, not all of these activities were controlled and monitored. Monitoring before mining or processing can be performed to establish baseline conditions. High-volume particulate matter (i.e., dust) or deposition monitoring is generally used to determine compliance with relevant air-quality standards and to ensure that the dust removal and suppression techniques are adequately working.

Targeted emissions monitoring can be performed to address site-specific issues, such as crystalline silica concentration, radon, or radioactivity. Other equipment is used to measure regulated pollutants such as sulfur oxides, which can affect downwind environments by dry or wet acid deposition. Zinc, arsenic, mercury, and cadmium will vaporize when heated during processing and can be released into the environment as gases that later precipitate or fall as dry deposition, sometimes many kilometers from the origin. All of these can affect water quality and soils downwind of the mining facilities. Special source sampling and monitoring programs should be developed to determine if these possible contaminants are issues. Sampling and monitoring of air quality, including various sampling procedures, sample collection devices, and real-time instruments, are summarized by McDermott (2004). After the World Trade Center collapse, the U.S. Geological Survey (USGS) examined the environmental effects of that disaster and summarized their results in Clark et al. (2001). Many of the techniques used by USGS can be employed at mine sites. For instance, scanning electron microscopy and X-ray microanalysis was useful in defining signature components of dust dispersed by the collapse of the World Trade Center (Lowers et al. 2005; Meeker et al. 2005).

level, or compliance with regulatory requirements. The constituent and type of air sampling used will depend on the objectives of the investigation. Commonly used types of air sampling include

- Personal exposure sampling,
- Area airborne sampling,
- Surface or underground sampling,
- Ambient sampling of criteria pollutants, and
- Emission sampling.

Personal exposure sampling is performed for occupational health and safety purposes. In personal exposure sampling, a sampling device is placed on an individual within his or her breathing zone. In area airborne sampling, sampling devices are employed to discern sources of contaminants or the effectiveness of cleanup procedures. Area samples should not be used as a substitute for personal exposure sampling. Surface sampling generally entails use of wipe samples to determine the presence of particulates or the effectiveness of equipment decontamination procedures. Underground sampling involves specialized equipment or monitors that can be connected to warning devices and ventilation systems for carbon monoxide and diesel emissions (Sengupta 1990). Ambient sampling involves collection of data over a wide area and includes fugitive dust (particulate matter) at property boundaries. These particles can be weighed for reporting the particulate mass and can also be submitted for elemental analysis. Emission sampling includes kiln, stack, or equipment sampling for milling and processing units that have regulated emission standards by agencies.

It is good practice to collect periodic samples of exhaust mine air. Some specific air-quality concerns exist at many mine sites (see Air Emissions sidebar). Metals present in respirable dust may pose health concerns (e.g., from blasting; see Figure 3.10). For example, respirable silica dust is typically produced when drilling, blasting, or cutting silica-containing rock. Exposure to respirable silica-bearing dust can result in silicosis, a potentially fatal lung disease (e.g., Chen et al. 2005). Potential issues for gases or vapors include radon daughter products, sulfur dioxide, mercury, hydrogen cyanide, methane, or hydrogen sulfide depending on the mine and processing type and activity. The use of diesel engines underground can produce a mixture of gases, vapors, and particulate matter including carbon monoxide, nitrogen dioxide, sulfur dioxide, volatile organic compounds, aldehydes, and polycyclic aromatic hydrocarbons. Diesel particulate matter is thought to be a probable human carcinogen (see USEPA 2003a).

3.8 QUALITY ASSURANCE/QUALITY CONTROL CONSIDERATIONS

QA/QC is a system of procedures, checks, audits, and corrective actions to ensure that samples are collected, stored, shipped, and analyzed in an acceptable manner with a stated level of confidence, and that technical aspects and reporting are adequate to ensure credible results (Taylor 1987). QA/QC is part of the overall USEPA Quality System and is generally described in the quality assurance project plan (QAPP). USEPA guidance states that the QAPP must specify the level of QA/QC activities necessary for the investigation and be composed of elements covering the entire project from planning, through implementation, to assessment (USEPA 2001b). The following four main groups of elements are included in a QAPP (USEPA 2001b):

1. **Project Management:** Project/task organization (roles and responsibilities), problem definition/background, project/task description, quality objectives and criteria, special training/certification, documents and records
2. **Data Generation and Acquisition:** Sampling design; sampling methods; sample preparation, handling, and custody; analytical methods; QC; instrument/equipment testing,

Courtesy of Virginia T. McLemore.

FIGURE 3.10 Blast at the Chino porphyry copper mine, Santa Rita district, New Mexico

inspection, and maintenance; instrument/equipment calibration and frequency; inspection/acceptance of supplies and consumables; nondirect measurements; data management

3. **Assessment and Oversight:** Assessments and response actions; reports to management
4. **Data Validation and Usability:** Data review, verification, and validation; verification and validation methods; reconciliation with user requirements

Traditionally, data quality concerns have largely concentrated on analytical errors. It is important to keep in mind that field sampling, sample preparation, and laboratory subsampling are generally the major sources of error in environmental investigations (Gy 1998; Ramsey and Hewitt 2005). Separate QA/QC procedures need to exist for all aspects of the sampling and monitoring program, including the field sampling program (collection, preservation, filtration, shipping, chain of custody, etc.), sample preparation (replicates at each stage of comminution), and the analytical procedures (laboratory). QA/QC is extremely important because "all analytical measurements are wrong: it's just a question of how large the errors are, and whether they are acceptable" (Thompson 1989).

As previously discussed, DQOs are qualitative and quantitative statements derived from systematic planning processes that clarify study objectives, define appropriate types of data, and specify tolerable levels of potential decision errors that will be used as the basis for establishing the quality and quantity of data needed to support decisions (USEPA 2006b). Data quality indicators (DQIs) are qualitative statistics and quantitative descriptors that are used to interpret the degree of acceptability or utility of data to the user and are part of the DQOs. The principal DQIs are precision, bias, representativeness, completeness, and comparability (USEPA 2006b), which are defined as follows (USEPA 2012).

- **Precision**—The degree to which a set of observations or measurements of the same property, usually obtained under similar conditions, conform to themselves; determined from multiple analysis of many sample replicates or standards.
- **Bias**—The systematic or persistent distortion of a measurement process.

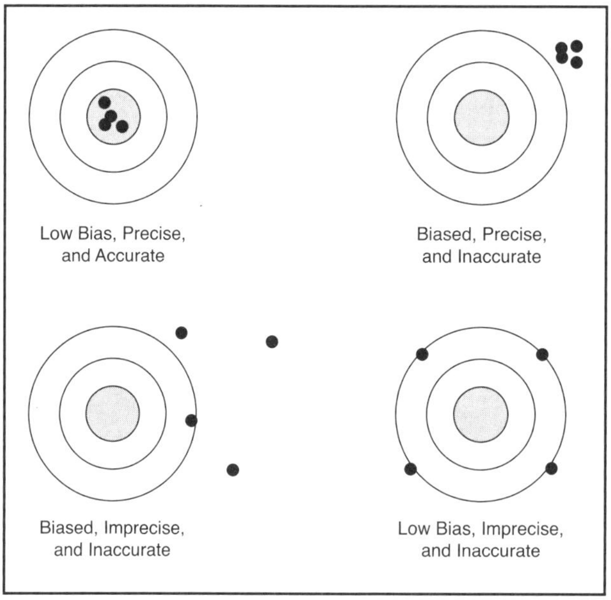

FIGURE 3.11 Diagram illustrating the difference between bias, precision, and accuracy for four measurements. The center of the target represents the true value and the dots represent individual measurements. For the bottom right target, note that individual measurements (dots) are inaccurate, but the mean of the four measurements is accurate.

- **Representativeness**—The extent to which measurements actually reflect the sampling unit from which they were taken, as well as the degree to which samples actually represent the target population at the time the samples were collected.
- **Completeness**—The amount of valid data obtained compared to the planned amount; usually expressed as a percentage.
- **Comparability**—The degree to which different methods, data sets, and/or decisions agree or can be represented as similar.

Myers (1997) defines the term *accuracy* as a summary of the closeness of measurements to the true (but possibly unknown) value. USEPA (2012) defines it as the degree of agreement between an observed value and an accepted reference value; determined by analyzing certified standards as if they were unknown samples and comparing those values with known certified values. *Precision*, *bias*, and *accuracy* should not be confused or used synonymously. Bias is usually a result of a systematic error (or errors). Figure 3.11 illustrates the classic target analogy for the distinction between bias, precision, and accuracy for repetitive analysis of one sample.

Environmental field measurements present challenges for estimating uncertainty. Qualitative and quantitative factors can contribute to field measurement uncertainty. Qualitative factors can include environmental conditions, sample handling, and instrument maintenance. Qualitative factors can be addressed through the preparation of standard operating procedures and the use of proper sampling equipment. Quantitative factors can include such things as instrument precision and accuracy, and calibration standards (USEPA 2008).

The goal of QC sampling is to identify, quantify, and document bias and variability in data that result from the collection, processing, shipping, and handling of samples (USGS, variously dated). The types of QC samples to be collected and their temporal and spatial frequency and distribution depend on the objectives of the investigation, data quality requirements, site conditions, and management or regulatory policy.

Accuracy can be evaluated by comparative sampling, performing analyses by alternate methods, or by analyzing certified reference materials using the same methods as are used in the investigation. Use of certified reference materials is an important part of a QA/QC program. An ARD-certified reference material can be obtained from the Canada Centre for Mineral and Energy Technology (CANMET) in Ottawa. Mine-waste reference materials also are available through the National Institute of Standards and Technology and USGS. Some Web sites for certified reference materials are listed in Appendix 1.

Generally, replicate samples or splits are used to determine precision for each stage of sample preparation in a sampling and monitoring program. Standards can also be analyzed repetitively to obtain a measure of precision. It is good practice and cost-effective to develop site-specific standard materials early in the sampling and monitoring program. Such standard materials need to be representative of the material being sampled in the sampling and monitoring plan. Multiple analyses of these standard materials, as well as analysis of sample replicates, can yield information about precision throughout the course of the investigation. Precision is usually expressed as a standard deviation, variance, or range in either absolute or relative terms. Evaluation of precision and accuracy needs to be built into every sampling and monitoring program.

Comparability can be addressed by using standardized sampling and analytical methods, and standardized units of reporting with similar sensitivity. For example, if one investigation collects a particular size fraction of waste material, the same fraction should be collected in a subsequent investigation to make the data comparable. It is always important to document any modifications of standard methods.

Blanks are used to assess possible contamination or inadvertent introduction of an analyte during the various stages of sampling and analysis. Types of blanks can include trip blanks (samples prepared prior to sampling that are not opened in the field but travel with the field samples), field blanks (samples prepared prior to sampling that are opened in the field during collection of a field sample), equipment blanks (water samples collected in the field from rinsing of sampling equipment), filter blanks (for water samples, clean water passed through filtration devices), and laboratory or method blanks (used to test for contamination from laboratory apparatus, sample preparation, and analysis).

Other types of QC samples can include matrix spikes, splits, and replicates. A matrix spike is a sample to which a known amount of the analyte is added to test for bias due to analytical interferences or matrix effects. Matrix effects refer to sample characteristics that interfere with the analytical method and can result in unreliable data. Splits are field and sample preparation samples that are split into two or more subsamples to allow for separate analysis to test laboratory performance. Laboratory splits of the prepared sample can be used when the same sample is being sent to different laboratories or to test different methods in the same laboratory. Replicates (or duplicates) are additional field samples collected at the same location under the same conditions with the same sampling protocol to test for sample variability and sampling error. The field replicate is one of the most important QC samples and should be collected every 10 to 20 samples. Replicates should be submitted *blind*; that is, with a designation that does not allow the laboratory to identify the original sample with which the replicate is associated. Table 3.2 summarizes information provided by some QC samples.

Methods of sample and data analysis need to be specified and pertinent literature referenced for standard, modified, and nonstandard methods. When using analytical techniques, each instrument needs to be calibrated with standard solutions prepared by the analyst or with QC

TABLE 3.2 Information provided by quality control samples

QC Type	Contamination					Precision			
	Containers	Field	Equipment	Cross-Contamination	Laboratory	Sampling	Splitting	Subsampling	Analysis
Blanks									
Trip	X			X	X				
Field	X	X		X	X				
Equipment	X	X	X	X	X				
Method					X				
Replicates and Splits									
Field splits							X	X	X
Field replicates						X		X	X
Lab splits							X	X	X
Lab replicates						X		X	X
Sample prep replicates							X	X	X

Source: Adapted from Wait and Ramsey 2007.

standards. Instrument calibration is carried out to measure sensitivity, determine resolution, and check linearity. The calibration data are maintained on file so as to monitor the performance of the instrument over time.

A chain of custody may be necessary as part of a QA/QC program. A chain of custody includes a record of all personnel who have handled the sample (with date, time, and place information) and documents all samples, fractions, splits, and sample disposal. This is also termed a *cradle to grave* record for a sample. Generally, chain of custody is used for legal purposes. However, it can be useful when trying to trace the source of anomalous data.

The QA/QC process does not end once the data are reported from the laboratory. Whenever possible, analytical results and laboratory documentation need to be independently reviewed for errors or omissions. Data verification is a process in which QC parameters are evaluated against a set of predetermined criteria or guidelines (e.g., DQIs). Data verification generally includes examination of holding times, calibrations, blanks, spikes, duplicates, laboratory control samples, and blind standards. It is common to have censored data when analyzing environmental samples. For example, environmental data commonly contain values that are below the detection limit (nondetects), and these values are generally reported as less than the limit of detection. Censored data pose problems when applying statistical methods to summarize and interpret data. Helsel (2005) provides guidance on dealing with censored data.

Table 3.3 is an example of a tool that could be used for data verification. The values listed are the generally accepted range for a number of parameters in water. If data fall outside the generally accepted range, it could indicate an error and the data should be flagged. These data should then be examined on a case-by-case basis during the data validation process. Several online sources of information about the natural abundance of elements are available that could be useful during the data verification phase. For example, WebElements.com (see Section 3.13) provides elemental

TABLE 3.3 High and low values used by the STORET database since November 1983 to screen or error-check data

Parameter	High Value	Low Value
pH (standard units)	12	0.9
Specific conductance (μmhos/cm)	60,000	1.0
Dissolved oxygen (mg/L)	30	0
Hardness, total (mg/L as $CaCO_3$)	5,000	0
Alkalinity, total (mg/L as $CaCO_3$)	1,000	0
Acidity, total (mg/L as $CaCO_3$)	1,000	0
Dissolved organic carbon (mg/L as C)	100	0
Calcium, dissolved (mg/L)	1,000	0
Magnesium, dissolved (mg/L)	1,000	0
Sodium, dissolved (mg/L)	5,000	0
Potassium, dissolved (mg/L)	175	0
Chloride, total in water (mg/L)	22,000	0
Sulfate, dissolved (mg/L)	2,500	0
Arsenic, dissolved (μg/L)	5,000	0
Barium, dissolved (μg/L)	2,000	0
Cadmium, dissolved (μg/L)	500	0
Copper, dissolved (μg/L)	2,000	0
Lead, dissolved (μg/L)	1,000	0
Manganese, dissolved (μg/L)	5,000	0
Mercury, dissolved (μg/L)	10	0
Nickel, dissolved (μg/L)	2,000	0
Zinc, dissolved (μg/L)	25,000	0

Source: Adapted from NPS 2001.

abundance information for a variety of media. Abundance information is also compiled by Smith and Huyck (1999) for different types of sampling media.

Data usability is determined during the data validation process and should be conducted by personnel familiar with the site, study objectives, and intended applications of the data. Data validation involves a detailed examination of the data using professional judgment to determine if the data are adequate to meet the needs of the investigation.

The DQO process is essential to the proper application of a QA/QC program. Francoeur (1997) discusses the consequences of having a *disconnect* between DQOs, QA/QC, and the verification of data quality. He describes *data time bombs* that result from this disconnect. In these cases, laboratory data go directly from the analytical instrument to data reports in which conclusions are drawn. He also points out that the mere performance of QC procedures does not ensure valid or usable data in the laboratory report and states that "the DQO process is the only certain means by which the data user is certain to generate data of sufficient quality to meet an intended purpose" (Francoeur 1997).

Because contaminants can occur at much higher levels at mine sites and because there can be physical hazards at mine sites, it is good practice to also develop a site health and safety plan prior to conducting field investigations. Personal protective equipment and specific training requirements for health and safety need to be considered at the outset of the investigation.

3.9 ANALYTICAL CHEMISTRY CONSIDERATIONS

Major causes of bias in environmental sampling and analysis include the following (USEPA 2006b):

- Nonrepresentative sampling
- Instability or contamination of samples between sampling and analysis
- Interferences and matrix effects in analysis
- Inability to determine the relevant forms of the parameter being measured
- Calibration
- Failure to blank-correct

It is not always appreciated how important sampling is in the analytical process. In many cases, sampling errors are the main contributor to the total error of the process (Gy 1998; Ramsey and Hewitt 2005), especially when trace constituents are being measured (see Section 3.4.8).

Selection of analytical methods is driven by the objectives of the investigation. Once the objectives have been identified, it is necessary to identify the appropriate sampling media and analytes of interest. Many federal and state regulations have established lists of parameters that may be required, but it is critical to also consider the ultimate application of the data. For example, if the objective of the investigation is to produce a risk assessment or provide data for input to a numerical model, then it is necessary to ensure that adequate data are collected and that the analytical detection limits are low enough to meet the requirements of these applications. Therefore, after establishing the media and analytes, selection of appropriate analytical methods needs to take into account the data needs of the investigation. It is important that personnel with adequate analytical chemistry expertise be involved in this planning phase of the investigation.

Appendix 3 provides a brief discussion of several analytical methods and media types. Some of the most popular references for analytical methods include *Standard Methods for the Examination of Water and Wastewater* (Rice et al. 2012), *Methods for Chemical Analysis of Water and Wastes* (USEPA 1983), *Test Methods for Evaluating Solid Waste (SW 846)* (USEPA, variously dated), and various ASTM methods (see also Appendix 2). The National Environmental Methods Index is an online database of method summaries that provides information to compare one method with another for the purpose of method selection, and can be used to search and compare regulatory and nonregulatory analytical methods (Keith et al. 2005; see Section 3.13). Some additional sources of online information are listed in Appendix 1. Compilations of methods can also be found in Dulski (1996), Crompton (2001), Csuros and Csuros (2002), Greenwood (2006), and Wagner (1998).

Deciding on appropriate analytical methods for an investigation involves several considerations, such as (1) analytes and detection limits needed; (2) suitability of sampling media for various analytical methods; (3) accuracy and precision needed; (4) limitations and interferences of various analytical instruments; (5) total or partial digestion, or direct analysis for solid materials; (6) need for nondestructive analysis; and (7) sample size limitations. Methods need to be sufficiently sensitive to reliably measure the concentration (i.e., adequate detection limits) and be free from serious interference problems. It is essential to ensure that detection limits are well below action levels and current regulatory guidelines for the contaminants of interest in the investigation. In the case of mineralized samples, upper detection limits also need to be considered when selecting analytical methods. It is essential that a person familiar with the data needs and applications, and the strengths and limitations of various analytical techniques, be involved in the choice of analytical methods for the investigation. It is also important to ensure that the user understands any method modifications or shortcuts employed by the laboratory.

Some other common analytical issues to consider include matrix effects and interferences from concomitant elements; volatility of constituents during sample collection, preparation, decomposition, or analysis; incomplete decomposition of sample when total analysis is desired; laboratory subsampling; and contamination problems. All analytical methods are potentially subject to interference by the physical or chemical constituents of the sample. Types of analytical interferences include spectral (for techniques such as inductively coupled plasma–atomic emission spectroscopy [ICP-AES]), physical matrix (which affects introduction of the sample into an analytical instrument), and space-charge effects (e.g., lighter elements are lost in transmission). The more complex the sample matrix, the more likely it is that interferences can cause significant analytical problems.

ICP-AES has become a commonly used analytical technique because of its multi-element capabilities (generally 30–40 elements), its large linear range (4 to 5 orders of magnitude; ppm to ppb), and its rapid throughput of samples (Crompton 2001). When lower limits of detection are needed, inductively coupled plasma–mass spectroscopy (ICP-MS) is often used. ICP-MS can determine 60+ elements and has a linear range of 6 to 7 orders of magnitude with detection limits at the ppb (parts per billion) to ppt (parts per trillion) level (see Wolf 2005 for additional information). A drawback of the ICP techniques is that they require liquid samples, so solid samples must undergo decomposition. If sample decomposition poses problems, an alternative technique is instrumental neutron activation analysis, which is a nondestructive technique that can determine 30 to 40 elements with a linear range of 5 orders of magnitude (ppm to ppb levels) (Crompton 2001). Laser ablation ICP-MS requires no digestion and has less interferences than most techniques. Its real strength is in spatial mapping of elements and their ratios in grains, inclusions, coatings, and so on (see USGS 2013a for additional information). Combustion-infrared methods are also widely used for carbon and sulfur mineral estimation methods. See Appendix 3 for additional information on these techniques.

Determining constituents in sampling media generally necessitates sample preparation (e.g., drying, disaggregation, sieving, crushing, grinding), subsampling for analysis, sample decomposition, and sample analysis. Water samples generally do not require preparation other than filtration and preservation. For solid samples, sample preparation considerations might include the need for staged comminution and subsampling to prepare a laboratory sample for assay, selection of a particular size fraction, segregation of minerals of interest, contamination from equipment, and loss of elements (e.g., loss of mercury if drying exceeds about 60°C [140°F]). Sample decomposition can be total, partial, or selective, depending on the requirements of the analytical technique and the purpose of the analysis. Sample decomposition can be problematic if an element(s) of interest resides in minerals that resist decomposition techniques. It is advisable to consider sample decomposition techniques when choosing analytical methods. Some sample preparation, decomposition, and analytical methods are discussed in Appendices 2 and 3.

The importance of sample preparation and laboratory subsampling warrants more attention than they have received. The effects of sample heterogeneity, sampling error, and sampling theory need to be considered in sample preparation and laboratory subsampling. Sample preparation subsampling errors can occur because the portion of a field sample that the laboratory analyzes from the prepared sample is typically *much* smaller than the field sample (Ramsey and Suggs 2001). When a sample preparation subsample is collected to prepare the laboratory sample, it is important that each subsample taken during sample preparation and laboratory analysis be representative of the field sample. For example, it is poor practice to scoop a few grains of a solid sample from the top of the sample container. This poor practice can result in significant errors because the concentration of the analyte at the top of the container is likely different from that at the bottom of the container. Gerlach et al. (2003) found that subsampling accuracy was at least 2 orders of magnitude worse than the accuracy of most analytical methods. USEPA provides guidance on laboratory subsampling (Gerlach et al. 2002; Gerlach and Nocerino 2003). For solid samples, crushing and grinding prior to subsampling can be an effective way to reduce sampling error (e.g., Walsh et al. 2002). Several devices have been designed to split samples (e.g., rotary sample dividers and riffle splitters), and the advantages and disadvantages of various designs are discussed by Schumacher et al. (1990) and Gerlach et al. (2002).

Cost-effectiveness, which differs from absolute cost, needs to be considered when choosing analytical methods. Often, a small increase in cost to optimize sampling, sample preparation, or sample analysis can greatly improve the overall effectiveness of the sampling and monitoring program. Conversely, an expensive technique needs to be evaluated as to its benefit to the program. Also, less expensive techniques can be considered for screening purposes at the beginning of the sampling program so that more samples can be analyzed (see Triad Resource Center, n.d.).

Selection of a qualified analytical laboratory is a critical step and needs to be conducted by personnel who are familiar with the investigation and the necessary analytical chemistry methods and QA/QC requirements. The laboratory should have the certifications (e.g., USEPA, State) necessary for the investigation, if permitting or compliance is involved, have a good reputation, have the analytical methods and detection limits appropriate for the investigation, be in a location where samples can be easily shipped or delivered, and be flexible enough to be able to meet any special requirements necessary for the investigation. Laboratory certification is not a substitute for data verification and data validation by personnel familiar with the site and the data applications. A laboratory should never be selected based solely on their analytical costs.

3.10 CONTAMINATION CONSIDERATIONS

During and after collection, the integrity of the sample needs to be maintained. It is nearly impossible to eliminate all forms of contamination. The most common causes of sample contamination during sample collection include poor sample-handling techniques, input from atmospheric sources, inadequately cleaned equipment, and use of equipment constructed of materials inappropriate for the analytes targeted for study (USGS, variously dated). Sample preservation and sample preparation methods and equipment can also contribute potential contaminants. In general, an awareness of possible routes of contamination, use of appropriate inert containers for the analytes of interest, and sufficient blank samples can help avoid many contamination problems. It is often necessary to use specially prepared or cleaned sampling containers. For example, DOC samples require glass containers that have been baked at high temperature to eliminate organic materials, and metals samples require inert containers that have been precleaned with acid. Numerous ASTM, federal, and state guidelines are available that specify proper containers for different kinds of media. It is important to investigate any pertinent requirements. Sample containers also need to be inert to the preservatives being used.

It is important to employ good field practices and clean sampling methods, and to routinely collect QC samples. Table 3.4 lists some common sources of contamination for water quality samples. Gasparon (1998) discusses ways to minimize contamination of water samples by trace metals. USGS (variously dated) lists the following activities that constitute good field practices for water quality sampling:

- Be aware of and record potential contamination sources at field sites.
- Wear appropriate, disposable, powderless gloves (ensure that gloved hands do not come in contact with the water sample, and change gloves before each new step of sample collection and processing).

TABLE 3.4 Common sources of contamination of water quality samples collected in the field

Contaminant Source Type	Examples
Sampling environment	Airborne particulates; precipitation; dust, soil, solid particles; fumes from engine exhaust, chemical preservatives, upwind industrial emissions
Sample-collection equipment	Pumps, isokinetic samplers, bailers, sample tubing, semipermeable membrane devices
Sample-processing equipment	Filtration devices, churn splitter, cone splitter, bottles, water (deionized/distilled water of ASTM grade 1 or better, tap, blank)
Sample-cleaning processes	Cleaning equipment (basins, brushes); carryover from cleaning solutions or tainted water; methanol carryover; insufficient decontamination or rinsing
Transport and shipping	Field vehicles; coolers or shipping containers; improperly closed or protected sample bottles
Storage	Warehouse; refrigerator; field vehicle; office laboratory; office storage space
Personnel	Dirty hands; sweat; sunscreen; DEET (active ingredient commonly used in insect repellents); nicotine, caffeine, and alcohol (breath); dirty gloves; gloved or ungloved contact with the sample to be analyzed; shedding clothing; hair and dandruff

Source: Adapted from USGS, variously dated.

- Use equipment constructed of materials that are relatively inert with respect to the analytes of interest.
- Use only equipment that has been precleaned according to prescribed procedures.
- Field-rinse equipment as directed.
- Use correct sample-handling procedures (minimize the number of sample-handling steps).
- Collect and process samples in enclosed chambers to minimize contamination from atmospheric sources.
- Collect sufficient blanks and other QC samples.
- Follow a prescribed order for collecting samples.

Human-induced contamination can originate from several sources such as sweat (which contains calcium, magnesium, lead, potassium, ammonium [NH_4^+], sulfate [SO_4^{2-}], phosphate [PO_4^{3-}], and cadmium for those who smoke), cosmetics (which can contain aluminum, beryllium, calcium, copper, chromium, potassium, iron, manganese, titanium, and zinc), eye makeup (which can contain mercury as a preservative), and dandruff shampoo (which can contain significant selenium) (see Gaines, n.d.-b).

Use of blanks can help identify contamination issues. It is good practice to prepare blanks for all steps of field sampling, sample handling, preservation, preparation, decomposition, and analysis. Specific types of blank samples are discussed in Section 3.8. It is advisable to sequence samples for analysis so that the blanks are analyzed first, which will prevent blanks from becoming contaminated by carryover contamination (i.e., samples with high concentrations contaminating those samples that follow) during analysis. Blank samples (e.g., pure deionized water) can also be used to test for carryover contamination by sequencing them throughout a set of samples. To be able to detect sample preparation and carryover contamination problems, it is necessary to know the order that samples were prepared and analyzed. If samples are randomly sequenced prior to preparation and analysis (a good practice), it is useful to first segregate them into groups expected to contain high concentrations versus low concentrations of analytes.

Use of chemical preservatives can be an important source of sample contamination. All sample preservatives need to be pre-approved by the analytical laboratory for the analytical techniques that will be used. It is good practice to use preservatives from the same lot for a given set of samples and record that information in the notes.

In collecting solid materials, contamination from sieves or other sampling apparatus needs to be considered. Also, contamination from a crushing and grinding apparatus can be an issue; common contaminants from tungsten carbide equipment can include tungsten, chromium, manganese, vanadium, cobalt, iron, copper, zirconium, and zinc (see Gaines, n.d.-a).

3.11 SAMPLE PRESERVATION, HANDLING, AND STORAGE CONSIDERATIONS

Sample preservation, handling, and storage can influence the integrity of the sample. Sample preservation is a very important consideration. Sample preservation by addition of chemical preservatives or refrigeration depends on the analyte and the analytical method to be used. All sample preservatives need to be pre-approved by the analytical laboratory for the analytical method and sample matrix. Sliwka-Kaszynska et al. (2003) provide a review of preservation and storage of water samples; Namiesnik and Szefer (2010) discuss considerations for preservation, storage, and analysis of water samples; and Biziuk et al. (2010) gives guidance about water-preservation methods.

Some issues that need to be considered during sampling, handling, and storage include

- Volatilization (and subsequent loss of analyte),
- Decomposition (agents can include temperature, ultraviolet radiation, microbial activity), and
- Chemical reactions (agents can include container walls, oxygen, carbon dioxide).

Some actions that can be taken to prevent these issues include

- Use of preservatives,
- Protecting samples from external agents, and
- Refrigeration or freezing.

For solid materials, analyte losses can occur during drying, crushing, grinding, decomposition, or heating, especially for volatile analytes. It is good practice to follow accepted preservation and storage protocols for various types of media. Appendix 3 provides discussions about different types of sampling media. Table 3.5 lists USEPA-recommended holding times and preservation methods for aqueous and solid samples. Some MIW samples may require special sample preservation procedures or shorter holding times.

3.12 SUMMARY

This chapter highlights information, resources, and research studies that relate to environmental sampling at mining and processing sites. Supplemental information relevant to this chapter can be found in Appendix 1 (online resources), Appendix 2 (ASTM methods), Appendix 3 (methods summaries), Appendix 5 (case studies), and Appendix 6 (geo-environmental models). This chapter is not a prescriptive guide for sampling at mining and processing sites. Rather, it points out selected sampling issues that need to be considered in the context of the sampling objectives for a given mining or processing site investigation. One of the most important aspects of sampling and monitoring is problem formulation (i.e., articulating the reason for sampling). This is a prevalent theme throughout this handbook and is also essential to the sampling considerations in this chapter. This chapter includes specialized sections about drillhole/borehole sampling considerations, biological (aquatic) sampling issues, microbial sampling considerations, and air sampling matters.

TABLE 3.5 USEPA-recommended sample holding times and preservation for selected inorganic analyte determinations in aqueous and solid samples

Analyte	Matrix	Fraction	Preservation	Holding Time
Metals, except Hg and Cr(VI)	Aqueous	Total	HNO_3 to pH<2	6 months
		Dissolved	Filter on site; HNO_3 to pH<2	6 months
	Solid	Suspended	Filter on site	6 months
		Total	None	6 months
Hexavalent chromium	Aqueous		≤6°C (43°F)	24 hours
	Solid		≤6°C (43°F)	30 days to extraction; 7 days from extraction to analysis
Mercury	Aqueous	Total	HNO_3 to pH<2	28 days
		Dissolved	Filter; HNO_3 to pH<2	28 days
	Solid	Total	≤6°C (43°F)	28 days
Chloride	Aqueous		≤6°C (43°F)	28 days
Cyanide	Aqueous		≤6°C (43°F); NaOH to pH>12	14 days
	Solid		≤6°C (43°F)	14 days
Fluoride	Aqueous		≤6°C (43°F)	28 days
Nitrate	Aqueous		≤6°C (43°F)	28 days
pH	Aqueous		NA	Analyze immediately
	Solid		NA	Analyze immediately
Specific conductance	Aqueous		NA	Analyze immediately
Sulfate	Aqueous		≤6°C (43°F)	28 days
Sulfide	Aqueous		4 drops 2 N zinc acetate/ 100 mL sample; NaOH to pH>9; minimize aeration; store headspace free at ≤6°C (43°F)	7 days
	Solid		Fill sample surface with 2 N zinc acetate until moistened; store headspace free at ≤6°C (43°F)	7 days
Organic carbon, total	Aqueous		≤6°C (43°F); store in dark; HCl or H_2SO_4 to pH<2	28 days
	Solid		≤6°C (43°F)	28 days

Source: Adapted from USEPA 2007e.

The sampling step for solid materials is often the main contributor to the total error in the overall process, especially when trace constituents are being measured. Many modern analytical techniques have errors <5%, whereas poor sampling or sample preparation can result in errors >100%. A sample needs to be properly collected and prepared to be representative of the target population, and the target population needs to be carefully defined based on the objectives of the investigation. Sample representativeness can be addressed by using a sampling theory approach. The major source of uncertainty in environmental data sets is due to the heterogeneity of environmental materials, and use of sampling theory can help minimize these sampling errors. Sampling theory is discussed in detail in this chapter.

Knowledge of the geology of an area is fundamental to planning for sampling programs. Frequently, one objective of a sampling program is to determine background, baseline, or premining conditions. GEMs are a geology-based classification system for identifying environmental challenges associated with extraction of specific types of mineral deposits and can provide an initial prediction of water quality at a given mine site based on the type of mineral deposit. GEMs are discussed in this chapter and also in Chapter 2 and Appendix 6. Mineralogical and petrographic characterization is also a source of fundamental information for solid materials and can lend important insights into potential for acid generation, neutralization, and contaminant release. Several aspects of mineralogical characterization are discussed in this chapter.

Some water-sampling issues covered in this chapter include filtration, unstable constituents, and parameters and constituents that need to be measured in the field or processing facility. Temporal variation of metal concentrations in surface water can be significant in streams draining mined and mineralized areas and in treated process water, and this needs to be considered in the context of the sampling objectives. Stream-water flow/discharge measurements can be invaluable when comparing solute concentrations between different sampling events and are necessary to calculate the load of constituents in water. Groundwater/surface-water interactions can be important and need to be considered in the overall sampling plan when pathways are being identified. For groundwater sampling, the importance of well design and purging methods are discussed. Water balances are fundamental to the conceptualization of hydrologic systems at all scales. Developing a water balance includes collection of hydrological (e.g., flow or discharge measurements, groundwater flow parameters) and climatological (e.g., precipitation, temperature, evapotranspiration) information.

QA/QC procedures need to exist for all aspects of the sampling and monitoring program, including the field sampling program (collection, preservation, filtration, shipping, chain of custody, etc.), sample preparation, and the analytical procedures (laboratory). Selection of analytical methods is driven by the objectives of the investigation. Once the objectives have been identified, it is necessary to identify the appropriate sampling media and analytes of interest. During and after collection, the integrity of the sample needs to be maintained. Sample preservation, handling, preparation, and storage can influence the integrity of the sample, and are all discussed in this chapter.

3.13 INTERNET RESOURCES

General Water Resources

National Water Quality Monitoring Council: http://acwi.gov/monitoring/
USEPA Field Sampling Procedures: www.epa.gov/region9/qa/fieldsamp.html
USEPA Water Home Page: http://water.epa.gov/
USGS Water Resources of the United States: http://water.usgs.gov/
USGS Water Quality Information Pages: http://water.usgs.gov/owq/
Water on the Web: www.waterontheweb.org/

Groundwater Resources

Global Groundwater Information System: www.igrac.net/
USEPA Ground Water and Drinking Water: http://water.epa.gov/drink/
USEPA Ground Water: http://water.epa.gov/type/groundwater/
USGS Groundwater Information Pages: http://water.usgs.gov/ogw/
USGS Water Resources Groundwater Software: http://water.usgs.gov/software/lists/groundwater/

Specialized Resources

ASTM International: www.astm.org/
Biotic Ligand Model (BLM) Research Version: www.hydroqual.com/wr_blm.html
EnviroStat's Definition of *Multi Increment*: www.envirostat.org/multiincrementdefinition.htm

National Environmental Methods Index: www.nemi.gov

Triad Resource Center: www.triadcentral.org

USEPA Agency-Wide Quality System Guidance Documents: www.epa.gov/quality/qa_docs.html

USEPA Decision Error Feasibility Trials (DEFT) software available for download: www.epa.gov/esd/databases/datahome.htm

USEPA Environmental Sampling: www2.epa.gov/region8/environmental-sampling

USEPA Soil Screening Guidance: www.epa.gov/superfund/health/conmedia/soil/index.htm

USGS Mineral Resources On-Line Spatial Data: http://mrdata.usgs.gov/

Visual Sample Plan (VSP) software tool available for download: http://vsp.pnnl.gov/

WebElements.com, Elemental Abundance Information: www.webelements.com/

Training

Contaminated-Site Clean-up Information (CLU-IN): www.clu-in.org

- Soil Sampling and Decision Making Using Incremental Sampling Methodology (recurring training)
- Hardrock Mining Geochemistry and Hydrology, Workshop 1 (archived at www.clu-in.org/conf/tio/r10hardrock/)
- Hardrock Mining Geochemistry and Hydrology, Workshop 2 (archived at www.clu-in.org/conf/tio/r10hardrock2/)
- Hardrock Mining Geochemistry and Hydrology, Workshop 3 (archived at www.clu-in.org/conf/tio/r10hardrock3_030513/)

EnviroStat: www.envirostat.org/training.htm

Interstate Technology and Regulatory Council (ITRC): www.itrcweb.org/ism-1/

Natural Water Quality Monitoring Council: http://acwi.gov/monitoring/webinars/

3.14 REFERENCES

AGDR (Australian Government Department of Resources). 2009. Airbourne contaminants, noise and vibration: Leading Practice Sustainable Development Program for the Mining Industry Report, www.ret.gov.au/resources/Documents/LPSDP/AirborneContaminantsNoiseVibrationHandbook_web.pdf. Accessed May 2013.

Al-Abed, S.R., Hageman, P.L., Jegadeesan, G., Madhavan, N., and Allen, D. 2006. Comparative evaluation of short-term leach tests for heavy metal release from mineral processing waste. *Sci. Total Environ.* 364:14–23.

Alaska Department of Environmental Conservation, Division of Spill Prevention and Response, Contaminated Sites Program. 2009. *Draft Guidance on Multi-Increment Soil Sampling.* www.dec.alaska.gov/spar/csp/guidance/multi_increment.pdf. Accessed May 2013.

Alaska DNR (Department of Natural Resources). 2013. Pebble project. Anchorage, AK: DNR. http://dnr.alaska.gov/mlw/mining/largemine/pebble/. Accessed 10/4/13.

Alley, W.M., Healy, R.W., LaBaugh, J.W., and Reilly, T.E. 2002. Flow and storage in groundwater systems. *Science* 296:1985–1990.

Alpers C.N., and Nordstrom, D.K. 1999. Geochemical modelling of water–rock interactions in mining environments. In *The Environmental Geochemistry of Mineral Deposits—Part A: Processes, Techniques, and Health Issues.* Reviews in Economic Geology, Vol. 6A. Littleton, CO: Society of Economic Geologists. pp. 289–324.

Alpers, C.N., and Nordstrom, D.K. 2000. Estimation of pre-mining conditions for trace metal mobility in mineralized areas—An overview. In *Proceedings, 5th International Conference on Acid Rock Drainage (ICARD 2000).* Littleton, CO: SME. pp. 463–472.

Amy, P.S., and Haldeman, D.L., eds. 1997. *The Microbiology of the Terrestrial Deep Subsurface.* New York: CRC Lewis.

Appelo, C.A.J., and Postma, D. 2005. *Geochemistry, Groundwater and Pollution,* 2nd ed. Leiden, The Netherlands: A.A. Balkema.

ASTM D 4448-01. 2007. *Standard Guide for Sampling Ground-Water Monitoring Wells.* West Conshohocken, PA: ASTM International. Available from www.astm.org.

ASTM D 4840-99. 2010. *Standard Guide for Sampling Chain-of-Custody Procedures*. West Conshohocken, PA: ASTM International. Available from www.astm.org.

ASTM D 6452-99. 2005. *Standard Guide for Purging Methods for Wells Used for Ground Water Quality Investigations*. West Conshohocken, PA: ASTM International. Available from www.astm.org.

Backhus, D.A., Ryan, J.N., Groher, D.M., McFarlane, J.K., and Gschwend, P.M. 1993. Sampling colloids and colloid-associated contaminants in ground water. *Ground Water*. 31(3):466–479.

Baedecker, M.J., and Cozzarelli, I.M. 1992. The determination and fate of unstable constituents of contaminated groundwater. In Lesage, Suzanne, and Jackson, R.E., eds., *Groundwater contamination and analysis at hazardous waste sites*. New York: Marcel Dekker. pp. 425–461.

Balistrieri, L.S., Seal, R.R., II, Piatak, N.M., and Paul, B. 2007. Assessing the concentration, speciation, and toxicity of dissolved metals during mixing of acid-mine drainage and ambient river water downstream of the Elizabeth Copper Mine, Vermont, USA. *Appl. Geochem*. 22:930–952.

Barbour, M.T., Gerritsen, J., Snyder, B.D., and Stribling, J.B. 1999. *Rapid Bioassessment Protocols for Use in Streams and Wadeable Rivers: Periphyton, Benthic Macroinvertebrates and Fish*, 2nd ed. EPA 841-B-99-002. Washington, DC: U.S. Environmental Protection Agency, Office of Water.

Barcelona, M.J., Gibb, J.P., Helfrich, J.A., and Garske, E.E. 1985. Practical Guide for Ground-Water Sampling. Champaign, IL: Illinois State Water Survey, SWS Contract Report 374 and U.S. Environmental Protection Agency EPA/600/S2-85/104. www.epa.gov/oust/cat/pracgw.pdf. Accessed May 2013.

Barcelona, M.J., Varlijen, M.D., Puls, R.W., and Kaminski, D. 2005. Ground water purging and sampling methods: History vs. hysteria. *Ground Water Monit. Remed*. 25:52–62.

BCMOE (British Columbia Ministry of Environment). 2012. Water and Air Baseline Monitoring Guidance Document for Mine Proponents and Operators, interim version. www.env.gov.bc.ca/epd/industrial/mining/pdf/water_air_baseline_monitoring.pdf. Accessed May 2013.

Bencala, K.E. 2000. Hyporheic zone hydrological processes: *Hydrol. Process*. 14:2797–2798.

Bencala, K.E. 2005. Hyporheic exchange flows. In *Encyclopedia of Hydrological Sciences*. Edited by M.G. Anderson. 3(10):1–7. New York: Wiley. http://water.usgs.gov/nrp/proj.bib/Publications/2005/bencala.2005.pdf. Accessed May 2013.

Bencala, K.E., Gooseff, M.N., and Kimball, B.A. 2011. Rethinking hyporheic flow and transient storage to advance understanding of stream-catchment connections. *Water Resources Res*. 47:W00H03.

Bervoets, L., Knaepkens, G., Eens, M., and R. Blust. 2005. Fish community responses to metal pollution. *Environ. Pollut*. 138:338–349.

Besser, J.M., and Leiby, K.J. 2007. Toxicity of metals in water and sediment to aquatic biota. In *Integrated Investigations of Environmental Effects of Historical Mining in the Animas River Watershed, San Juan County, Colorado*. Edited by S.E. Church, P. von Guerard, and S.E. Finger. Professional Paper 1651. Reston, VA: U.S. Geological Survey. http://pubs.usgs.gov/pp/1651/. Accessed May 2013.

Besser, J.M., Brumbaugh, W.G., May, T.W., Church, S.E., and Kimball, B.A. 2001. Bioavailability of metals in stream food webs and hazards to brook trout (*Salvelinus fontinalis*) in the upper Animas River Watershed, Colorado. *Arch. Environ. Contam. Toxicol.* 40:48–59.

Bish, D.L., and Howard, S.A. 1988. Quantitative phase analysis using the Rietveld method. *J. Appl. Crystallogr*. 21:86–91.

Bish, D.L., and Post, J.E., eds. 1989. *Modern Powder Diffraction*. Reviews in Mineralogy, Vol. 20. Washington, DC: Mineralogical Society of America.

Bish, D.L. and Post, J.E. 1993. Quantitative mineralogical analysis using the Rietveld full-pattern fitting method: *Am. Mineral*. 78:932–940.

Biziuk, M., Beyer, A., and Zukowska, J. 2010. Preservation and storage of water samples. In *Analytical Measurements in Aquatic Environments*. Edited by J. Namiesnik and P. Szefer. Taylor and Francis Group, CRC Press. pp. 19–39.

Borrok, D.M., Wanty, R.B., Ridley, W.I., Lamothe, P.J., Kimball, B.A., Verplanck, P.L., and Runkel, R.L. 2009. Distinguishing between natural and anthropogenic metal sources in a mountain watershed using iron and zinc isotopes. *Appl. Geochem*. 24:1270–1277.

Bove, D.J., Mast, M.A., Dalton, J.B., Wright, W.G., and Yager, D.B. 2007a. Major styles of mineralization and hydrothermal alteration and related solid- and aqueous-phase geochemical signatures. In *Integrated Investigations of Environmental Effects of Historical Mining in the Animas River Watershed, San Juan County, Colorado*. Edited by S.E. Church, P. von Guerard, and S.E. Finger. Professional Paper 1651. Reston, VA: U.S. Geological Survey. http://pubs.usgs.gov/pp/1651/. Accessed May 2013.

Bove, D.J., Yager, D.B., Mast, M.A., and Dalton, J.B. 2007b. *Alteration Map Showing Major Faults and Veins and Associated Water-Quality Signatures of the Animas River Watershed Headwaters near Silverton, Southwest Colorado*: Scientific Investigations Map 2976. pamphlet, 1 plate, scale 1:24,000. Reston, VA: U.S. Geological Survey.

Boyer, E.W., Hornberger, G.M., Bencala, K.E., and McKnight, D.M. 1997. Response characteristics of DOC flushing in an alpine catchment. *Hydrol. Processes*. 11:1635–1647.

Brumbaugh, W.G., Petty, J.D., Huckins, J.N., and Manahan, S.E. 2002. Stabilized liquid membrane device (SLMD) for the passive, integrative sampling of labile metals in water. *Water, Air, Soil Pollut.* 133:109–119.

Cabri, L.J., and Vaughan, D.J., eds. 1998. *Modern Approaches to Ore and Environmental Mineralogy*. Short Course Series, Vol. 27. Ottawa, Ontario: Mineralogical Association of Canada.

Callender, E. 2003. Heavy metals in the environment—Historical trends. *Treatise on Geochemistry*. 9:67–105.

Carlisle, D.M., and Clements, W.H. 1999. Sensitivity and variability of metrics used in biological assessment of running waters. *Environ. Toxicol. Chem*. 18(2):285–291.

Carlisle, D.M., Hawkins, C.P., Meador, M.R., Potapova, M., and Falcone, J. 2008. Biological assessment of Appalachian streams based on predictive models for fish, macroinvertebrate, and diatom assemblages. *J. N. Am. Benthol. Soc.* 27(1):16–37.

Carter, M.R., and Gregorich, E.G., eds. 2008. *Soil Sampling and Methods of Analysis*, 2nd ed. Canadian Society of Soil Science, Taylor and Francis Group, CRC Press.

Caruso, B.S., Cox, T.J., Runkel, R.L., Velleux, M.L., Bencala, K.E., Nordstrom, D.K., Julien, P.Y., Butler, B.A., Alpers, C.N., Marion, A., and Smith, K.S. 2008. Metals fate and transport modelling in streams and watersheds: State of the science and USEPA workshop review. *Hydrol. Processes.* 22:4011–4021.

Castendyk, D.N., and Eary, L.E., eds. 2009. *Mine Pit Lakes: Characteristics, Predictive Modeling, and Sustainability*. Management Technologies for Metal Mining Influenced Water series, Vol. 3. Littleton, CO: SME.

CDPHE (Colorado Department of Public Health and Environment). 2011. *Clear Creek*: Hazardous Materials and Waste Management Division, Central City/Clear Creek. www.colorado.gov/cs/Satellite/CDPHE-HM/CBON/1251615994932. Accessed May 2013.

Chapelle, F.H. 1993. *Ground-Water Microbiology and Geochemistry*. New York: John Wiley and Sons.

Chapin, T.P., and Todd, A.S. 2012. MiniSipper: A new in situ water sampler for high-resolution, long-duration acid mine drainage monitoring. *Sci. Total Environ*. 439:343–353.

Chen, W., Henizdo, E., Chen, J.-Q., Attfield, M.D., Gao, P., Hearl, F., Lu, J., and Wallace, W.E. 2005. Risk of silicosis in cohorts of Chinese tin and tungsten miners, and pottery workers (I): An epidemiological study. *Am. J. Ind. Med*. 48:1–9.

Choate, L.M., Ranville, J.F., Blumenstein, E.P., and Smith, K.S. 2008. Simple screening tests to determine the potential metal toxicity of mining-waste leachates. In *Understanding Contaminants Associated with Mineral Deposits*. Edited by P.L. Verplanck. Circular 1328, pp. 24–27. Reston, VA: U.S. Geological Survey. http://pubs.usgs.gov/circ/1328/. Accessed May 2013.

Church, P.E., and Granato, G.E. 1996. Bias in ground-water data caused by well-bore flow in long-screen wells. *Ground Water* 34:262-273.

Church, S.E., Fey, D.L., and Unruh, D.M. 2007a. Trace elements and lead isotopes in modern streambed and terrace sediment—Determination of current and premining geochemical baselines. In *Integrated Investigations of Environmental Effects of Historical Mining in the Animas River Watershed, San Juan County, Colorado*. Edited by S.E. Church, P. von Guerard, and S.E. Finger. Professional Paper 1651. Reston, VA: U.S. Geological Survey. http://pubs.usgs.gov/pp/1651/. Accessed May 2013.

Church, S.E., Kirschner, F.K., Choate, L.M., Lamothe, P.J., Budahn, J.R., and Brown, Z.A. 2007b. *Determination of Premining Geochemical Background, Blue Creek Drainage, Stevens County, Washington*. Scientific Investigations Report 2007-5262. Reston, VA: U.S. Geological Survey. http://pubs.usgs.gov/sir/2007/5262/. Accessed May 2013.

Church, S.E., von Guerard, P., and Finger, S.E., eds. 2007c. *Integrated Investigations of Environmental Effects of Historical Mining in the Animas River Watershed, San Juan County, Colorado*. Professional Paper 1651. Reston, VA: U.S. Geological Survey. http://pubs.usgs.gov/pp/1651/. Accessed May 2013.

Clark, J.L., and Clements, W.H. 2006. The use of in situ and stream microcosm experiments to assess population- and community-level responses to metals. *Environ. Toxicol. Chem*. 25:2306–2312.

Clark, R.N., Green, R.O., Swayze, G.A., Meeker, G., Sutley, S., Hoefen, T.M., Livo, K.E., Plumlee, G., Pavri, B., Sarture, C., Wilson, S., Hagemen, P., Lamothe, P., Vance, J.S., Boardman, J., Brownfield, I., Gent, C., Morath, L.C., Taggart, J., Theodorakos, P.M., and Adams, M. 2001. *Environmental studies of the World Trade Center area after the September 11, 2001 attack.* U.S. Geological Survey, Open-File Report 01-0429, http://pubs.usgs.gov/of/2001/ofr-01-0429/. Accessed May 2013.

Clements, W.H., and Kiffney, P.M. 1995. The influence of elevation on benthic community responses to heavy metals in Rocky Mountain streams. *Can. J. Fish. Aq. Sci.* 542:1966–1977.

Clements, W.H., Carlisle, D.M., Lazorchak, J.M., and Johnson, P.C. 2000. Heavy metals structure benthic communities in Colorado mountain streams: *Ecol. Appl.* 10:626–638.

Cochran, W.G. 1977. *Sampling Techniques*, 3rd ed. New York: John Wiley and Sons.

Cohen, D.R., and Ward, C.R. 1991. SEDNORM—A program to calculate a normative mineralogy for sedimentary rocks based on chemical analyses. *Computers Geosci.* 17:1235–1253.

Craig, J.R., and Vaughan, D.J. 1994. *Ore Microscopy and Ore Petrography*, 2nd ed. New York: John Wiley and Sons.

Cravotta, C.A., III. 1994. Secondary iron-sulfate minerals as sources of sulfate and acidity. In *Environmental Geochemistry of Sulfide Oxidation*, ACS Symposium Series, Vol. 550. Edited by C.N. Alpers and D.W. Blowes. Washington, DC: American Chemical Society.

Crompton, T.R. 2001. *Determination of Metals and Anions in Soils, Sediments, and Sludges*. London: Taylor and Francis Group, Spon Press.

Crumbling, D.M., Lynch, K., Howe, R., Groenjes, C., Shockley, J., Keith, L., Lesnik, B., Van E, J., and McKenna, J. 2001. Managing uncertainty in environmental decisions. *Environ. Sci. Technol.* 35:404A–409A.

Csuros, M., and Csuros, C. 2002. *Environmental Sampling and Analysis for Metals*. Boca Raton, FL: CRC Press, Lewis Publishers.

Custer, C.M, Yang, C., Crock, J.G., Shearn-Bochsler, V., Smith, K.S., and Hageman, P.L. 2009. Exposure of insects and insectivorous birds to metals and other elements from abandoned mine tailings in three Summit County drainages, Colorado. *Environ. Monit. Assess.* 153:161–177.

CWQCC (Colorado Water Quality Control Commission). 2010. *Aquatic Life Use Attainment—Methodology to Determine Use Attainment for Rivers and Streams*, Policy Statement 10-1, October 2010. Colorado Department of Public Health and Environment. www.colorado.gov/cs/Satellite?blobcol=urldata&blobheadername1=Content-Disposition&blobheadername2=Content-Type&blobheadervalue1=inline%3B+filename%3D%22Policy+10-1.pdf%22&blobheadervalue2=application%2Fpdf&blobkey=id&blobtable=MungoBlobs&blobwhere=1251853603204&ssbinary=true. Accessed May 2013.

Danielsson, L.G. 1982. On the use of filters for distinguishing between dissolved and particulate fractions in natural waters. *Water Res.* 16:179–182.

Degens, E.T., Kempe, S., and Richey, J.E., eds. 1991. *Biogeochemistry of Major World Rivers*, SCOPE Vol. 42. International Council of Scientific Unions, Scientific Committee on Problems of the Environment (SCOPE). www.scopenvironment.org/downloadpubs/scope42/index.html. Accessed May 2013.

Delvigne, J.E. 1998. *Atlas of Micromorphology of Mineral Alteration and Weathering*, Special Publication No. 3. Canadian Mineralogist.

Desborough, G.A., Smith, K.S., Lowers, H.A., Swayze, G.A., Hammarstrom, J.M., Diehl, S.F., Leinz, R.W., and Driscoll, R.L. 2010. Mineralogical and chemical characteristics of some natural jarosites. *Geochim. Cosmochim. Acta* 74:1041–1056.

Diehl, S.F., Hageman, P.L., and Smith, K.S. 2006. What's weathering? Mineralogy and field leach studies in mine waste, Leadville and Montezuma mining districts, Colorado. In *Proceedings, 7th International Conference on Acid Rock Drainage (ICARD 7)*. Lexington, KY: American Society of Mining and Reclamation. pp. 507–527. www.imwa.info/docs/imwa_2006/0507-Diehl-CO.pdf. Accessed May 2013.

Diehl, S.F., Koenig, A.E., Hageman, P.L., Smith, K.S., Fey, D.L., and Lowers, H.A. 2007. From the micro to the macro scale: A textural and chemical perspective of characterizing waste-rock material. In *Proceedings, Society for Mining, Metallurgy, and Exploration (SME) Annual Meeting and Exhibit, and the 109th National Western Mining Conference*, Preprint 07-021. Littleton, CO: SME.

Diehl, S.A., Hageman, P.L., and Smith, K.S. 2008. What is weathering in mine waste? Mineralogic evidence for sources of metals in leachates. In *Understanding Contaminants Associated with Mineral Deposits*. Edited by P.L. Verplanck. Circular 1328. Reston, VA: U.S. Geological Survey. http://pubs.usgs.gov/circ/1328/. Accessed May 2013.

Diehl, S.F., Hageman, P.L., Koenig, A.E., Seal, R.R., II, Smith, K.S., Benzel, W.M., and Driscoll, R.L. 2010. Techniques to understand metal leaching from ore minerals in mine-waste material. In *Proceedings, SME Annual Meeting and Exhibit*. Littleton, CO: SME.

Di Toro, D.M., Allen, H.E., Bergman, H.L, Meyer, J.S., Paquin, P.R., and Santore, R.C. 2001. A biotic ligand model of the acute toxicity of metals. I. Technical basis. *Environ. Toxicol. Chem.* 20:2383–2396.

Downing, B. 2011. ARD sampling and sample preparation: *EnviroMine*. http://technology.infomine.com/enviromine/ard/sampling/intro.html. Accessed May 2013.

Drexler, J.W., and Brattin, W.J. 2007. An in vitro procedure for estimation of lead relative bioavailability: with validation: *Human Ecol. Risk Assess.* 13:383–401.

Druschel, G.K., Baker, B.J., Gihring, T.M., and Banfield, J.F. 2004. Acid mine drainage biogeochemistry at Iron Mountain, California. *Geochem. Trans.* 5(2):13–31.

du Bray, E.A., ed. 1995. *Preliminary Compilation of Descriptive Geoenvironmental Mineral Deposit Models*. Open-File Report 95-0831. Reston, VA: U.S. Geological Survey. http://pubs.usgs.gov/of/1995/ofr-95-0831/. Accessed May 2013.

Dulski, T.R. 1996. *A Manual for the Chemical Analysis of Metals*. West Conshohocken, PA: ASTM International.

Eary, L.E., Jakubowski, R., Eshleman, J., and Watson, A. 2009. Applying numerical hydrogeochemical models as decision support tools for mine closure planning. In *Proceedings, Tailings and Mine Waste '08*. London: Taylor and Francis Group, CRC Press/Balkema.

ECA (European Chemicals Agency). 2008. *Guidance on information requirements and chemical safety assessment—Appendix R.7.13-2: Environmental risk assessment for metals and metal compounds*: European Chemicals Agency, Guidance for the implementation of REACH. http://echa.europa.eu/documents/10162/13632/information_requirements_r7_13_2_en.pdf. Accessed June 2013.

Edwards, T.K., and Glysson, G.D. 1999. Field methods for measurement of fluvial sediment. *Techniques of Water-Resources Investigations*, Book 3. Reston, VA: U.S. Geological Survey. http://pubs.usgs.gov/twri/twri3-c2/. Accessed May 2013.

Environment Canada. 2009. *Environmental Code of Practice for Metal Mines*. www.ec.gc.ca/lcpe-cepa/default.asp?lang=En&n=CBE3CD59-1. Accessed May 2013.

EnviroStat. n.d. Definition of *Multi Increment*. www.envirostat.org/multiincrementdefinition.htm. Accessed 12/15/13.

Farag, A.M., Skaar, D., Nimick, D.A., MacConnell, E., and C. Hogstrand, C. 2003. Characterizing aquatic health using salmonid mortality, physiology, and biomass estimates in streams with elevated concentrations of arsenic, cadmium, copper, lead and zinc in the Boulder River watershed, Montana. *Trans. Am. Fish. Soc.* 132:450–467.

Fenter, P.A., Rivers, M.L., Sturchio, N.C., and Sutton, S.R., eds. 2002. *Applications of Synchrotron Radiation in Low-Temperature Geochemistry and Environmental Science*. Reviews in Mineralogy and Geochemistry, Vol. 49. Washington, DC: Mineralogical Society of America.

Fetter, C.W. 2000. *Applied Hydrogeology*, 4th ed. Upper Saddle River, NJ: Prentice Hall.

Ficklin, W.H., and Mosier, E.L. 1999. Field methods for sampling and analysis of environmental samples for unstable and selected stable constituents. In *The Environmental Geochemistry of Mineral Deposits—Part A: Processes, Techniques, and Health Issues*. Reviews in Economic Geology, Vol. 6A. Littleton, CO: Society of Economic Geologists. pp. 249–264.

Fleet, M.E. 2005. XANES spectroscopy of sulfur in earth materials. *Can. Mineral.* 43:1811–1838.

Foster A.L., Brown, G.E., Jr., Tingle, T., and Parks, G.A. 1998. Quantitative arsenic speciation in mine tailings using X-ray absorption spectroscopy. *Am Mineral.* 83:553–568.

Francoeur, T. L. 1997. Quality control: The great myth. In *Proceedings of the Field Analytical Methods for Hazardous Wastes and Toxic Chemicals Conference*, January 29–31, 1997, Las Vegas, NV. Pittsburgh, PA: Air and Waste Management Association. pp. 651–657.

Freeze, R.A., and Cherry, J.A. 1979. *Groundwater*. Upper Saddle River, NJ: Prentice Hall.

Fuller, C.C. and Harvey, J.W. 2000. Reactive uptake of trace metals in the hyporheic zone of a mining-contaminated stream, Pinal Creek, Arizona. *Environ. Sci. Technol.* 34:1150–1155.

Gaines, P. n.d.-a. Part 3: Sampling and subsampling. In *Trace Analysis Guide*. Web series. Christiansburg, VA: Inorganic Ventures. www.inorganicventures.com/sampling-and-subsampling. Accessed 10/4/13.

Gaines, P. n.d.-b. Part 10: Contamination from the analyst and apparatus. In *Trace Analysis Guide*. Web series. Christiansburg, VA: Inorganic Ventures. www.inorganicventures.com/tech/trace-analysis/contamination-from-the-analyst-and-apparatus. Accessed 10/4/13.

Gałuszka, A. 2007. A review of geochemical background concepts and an example using data from Poland. *Environ. Geol.* 52(5):861–870.

Gammons, C.H., Nimick, D.A., Parker, S.R., Cleasby, T.E., and McCleskey, R.B. 2005. Diel behavior of iron and other trace metals in a mountain stream with acidic to neutral pH: Fisher Creek, Montana, USA. *Geochim. Cosmochim. Acta* 69:2505–2516.

Gammons, C.H., Milodragovich, L., and Belanger-Woods, J. 2007. Influence of diurnal cycles on monitoring of metal concentrations and loads in streams draining abandoned mine lands: An example from High Ore Creek, Montana: *Environ. Geol.* 53:611–622.

Gandy, C.J., Smith, J.W.N., and Jarvis, A.P. 2007. Attenuation of mining derived pollutants in the hyporheic zone: A review: *Sci. Total Environ.* 373:435–446.

Gasparon, M. 1998. Trace metals in water samples: Minimizing contamination during sampling and storage. *Environ. Geol.* 36:207–214.

Gerlach, R.W., and Nocerino, J.M. 2003. *Guidance for Obtaining Representative Laboratory Analytical Subsamples from Particulate Laboratory Samples.* EPA/600/R-03/027. Washington, DC: U.S. Environmental Protection Agency. www.epa.gov/osp/hstl/tsc/Gerlach2003.pdf. Accessed May 2013.

Gerlach, R.W., Dobb, D.E., Raab, G.A., and Nocerino, J.M. 2002. Gy sampling theory in environmental studies. 1. Assessing soil splitting protocols. *J. Chemometr.* 16:321–328.

Gerlach, R.W., Nocerino, J.M., Ramsey, C.A., and Venner, B.C. 2003. Gy sampling theory in environmental studies. 2. Subsampling error estimates. *Anal. Chim. Acta* 490:159–168.

Gilbert, R.O. 1987. *Statistical Methods for Environmental Pollution Monitoring.* New York: John Wiley and Sons.

Gilbert, R.O., and Pulsipher, B.A. 2005. Role of sampling designs in obtaining representative data. *Environ. Forensics.* 6:27–33.

Glasgow, H.B., Burkholder, J.M., Reed, R.E., Lewitus, A.J., and Kleinman, J.E. 2004. Real-time remote monitoring of water quality: A review of current applications and advancements in sensor, telemetry, and computing technologies: *J. Exper. Mar. Biol. Ecol.* 300:409–448.

Goldstein, J.N., Woodward, D.F., and Farag, A.M. 1999. Movements of adult Chinook salmon during spawning migration in a metals-contaminated system, Couer d'Alene River, Idaho. *Trans. Am. Fish. Soc.* 128:121–129.

Greenwood, R. 2006. Directory of "screening tools"—A toolbox of existing and emerging methods for chemical and ecological status monitoring under the Water Framework Directive. In *Screening Methods for Water Data Information in Support of the Implementation of the Water Framework Directive*, Deliverable D5.

GRI (Global Reporting Initiative). 2010. *The GRI Mining and Metals Sector Supplement.* Global Reporting Initiative. www.icmm.com/page/36353/gri-mining-and-metals-sector-supplement. Accessed May 2013.

Griffith, M.B., Lazorchak, J.M., and Herlihy, A.T. 2004. Relationships among exceedances of metals criteria, the results of ambient bioassays, and community metrics in mining-impacted streams. *Environ. Toxicol. Chem.* 23:1786–1795.

Gschwend, P.M., and Reynolds, M.D. 1987. Monodisperse ferrous phosphate colloids in an anoxic groundwater plume. *J. Contam. Hydrol.* 1:309–327.

Gusek, J.J., and Figueroa, L.A., eds. 2009. *Mitigation of Metal Mining Influenced Water.* Management Technologies for Metal Mining Influenced Water series, Vol. 2. Littleton, CO: SME.

Gy, P. 1982. *Sampling of Particulate Materials: Theory and Practice*, 2nd ed. New York: Elsevier Science and Technology.

Gy, P. 1998. *Sampling for Analytical Purposes.* New York: John Wiley and Sons.

Hageman, P.L. 2007. *U.S. Geological Survey Field Leach Test for Assessing Water Reactivity and Leaching Potential of Mine Wastes, Soils, and Other Geologic and Environmental Materials.* Techniques and Methods, book 5. Reston, VA: U.S. Geological Survey. http://pubs.usgs.gov/tm/2007/05D03/pdf/TM5-D3_508.pdf. Accessed May 2013.

Hageman, P.L., and Briggs, P.H. 2000. A simple field leach for rapid screening and qualitative characterization of mine waste material on abandoned mine lands. In *Proceedings, 5th International Conference on Acid Rock Drainage (ICARD 2000)*. Littleton, CO: SME. pp. 1463–1475. http://crustal.usgs.gov/projects/minewaste/pdfs/hageman1.pdf. Accessed May 2013.

Hagni, A.M. 2008. Phase identification, phase quantification, and phase association determinations utilizing automated mineralogy technology. *J. Min. Met. Mat. Soc.* 60(4):33–37.

Haimes, Y.Y. 1998. *Risk Modeling, Assessment, and Management*. New York: John Wiley and Sons.

Hall, J.S. 2004. New Mexico Bureau of Mines and Mineral Resource's clay laboratory manual. Unpublished report. Socorro, NM: New Mexico Bureau of Geology and Mineral Resources.

Hamilton, S.J., Mehrle, P.M., and Jones, J.R. 1987. Evaluation of metallothionein measurement as a biological indicator of stress from cadmium in brook trout. *Trans. Am. Fish. Soc.* 116:551–560.

Hammarstrom, J.M., and Smith, K.S. 2002. Geochemical and mineralogical characterization of solids and their effects on waters in metal-mining environments. In *Progress on Geoenvironmental Models for Selected Mineral Deposit Types*. Edited by R.R. Seal II, and N.K. Foley. Open-File Report 02-0195. Reston, VA: U.S. Geological Survey. pp. 8–54. http://pubs.usgs.gov/of/2002/of02-195/. Accessed May 2013.

Harper, D.D., Farag, A.M., and Brumbaugh, W.G. 2008. Effects of acclimation on the toxicity of stream water contaminated with zinc and cadmium to juvenile cutthroat trout. *Arch. Environ. Contam. Toxicol.* 54:697–704.

Harvey, J.W., and Bencala, K.E. 1993. The effect of streambed topography on surface-subsurface water exchange in mountain catchments. *Water Resour. Res.* 29:89–98.

Harvey, J.W., and Fuller, C.C. 1998. Effect of enhanced manganese oxidation in the hyporheic zone on basin-scale geochemical mass balance. *Water Resour. Res.* 34:623–636.

Hauff, P.L., Peters, D.C., Coulter, D.W., Sares, M.A., Bird, D.A., Henderson, F.B., III, and Prosh, E.C. 2006. Hyperspectral sensing of acid mine drainage—Two Colorado case studies. In *Proceedings, 7th International Conference on Acid Rock Drainage (ICARD 7)*. Lexington, KY: American Society of Mining and Reclamation. pp. 738–763. www.imwa.info/docs/imwa_2006/0738-Hauff-CO.pdf. Accessed May 2013.

Hawaii Department of Health. 2009. Section 4.2: Multi-Increment sample collection. In Technical Guidance Manual for the Implementation of the Hawai'i State Contingency Plan. www.hawaiidoh.org/tgm.aspx?p=0402a.aspx. Accessed 10/4/13.

Hawkins, C.P., Norris, R.H., Hogue, J.N., and Feminella, J.W. 2000. Development and evaluation of predictive models for measuring the biological integrity of streams. *Ecol. Appl.* 10:1456–1477.

Healy, R.W., and Scanlon, B.R. 2010. *Estimating Groundwater Recharge*. New York: Cambridge University Press.

Healy, R.W., Winter, T.C., LaBaugh, J.W., and Franke, O.L. 2007. *Water Budgets: Foundations for Effective Water-Resources and Environmental Management*. Circular 1308. Reston, VA: U.S. Geological Survey. http://pubs.usgs.gov/circ/2007/1308/. Accessed May 2013.

Helsel, D.R. 2005. *Nondetects and Data Analysis, Statistics for Censored Environmental Data*. New York: John Wiley and Sons.

Hem, J. D. 1985. *Study and Interpretation of the Chemical Characteristics of Natural Water*, 3rd ed. Water Supply Paper 2254. Reston, VA: U.S. Geological Survey. http://pubs.usgs.gov/wsp/wsp2254/. Accessed May 2013.

Hoal, K.O., Smith, K.S., Russell, C.C., Walton-Day, K., and Stammer, J.G. 2009a. Application of quantitative micro-mineralogy to tailings and mining waste. In *Proceedings, 13th International Conference on Tailings and Mine Waste (Tailings & Mine Waste '09)*, Banff, Alberta, Canada, November 1-4, 2009. Edited by D. Sego, M. Alostaz, and N. Beier. Alberta, Canada: University of Alberta, Department of Civil and Environmental Engineering. pp. 703–709.

Hoal, K.O., Stammer, J.G., Appleby, S.K., Botha, J., Ross, J.K., and Botha, P.W. 2009b. Research in quantitative mineralogy: Examples from diverse applications. *Miner. Eng.* 22:402–408.

Hoar, C.C. 2005. *Fish Response to Discharge Events from a Power Plant Cooling Reservoir in a River Affected by Acid Mine Drainage and Thermal Influences*. West Virginia University M.S. thesis.

Hochella, M.F. 1995. Mineral surfaces: Their characterization and their chemical, physical and reactive nature. In *Mineral Surfaces*, Mineralogical Society, Series 5. Edited by D.J. Vaughan and R.A.D. Pattrick. London: Chapman and Hall. pp. 17–60.

Horowitz, A.J. 2008. Determining annual suspended sediment and sediment-associated trace element and nutrient fluxes. *Sci. Total Environ.* 400:315–343.

Horowitz, A.J., Elrick, K.A., and Colberg, M.R. 1992. The effect of membrane filtration artifacts on dissolved trace element concentrations. *Water Res.* 26:753–763.

Horowitz, A.J., Demas, C.R., Fitzgerald, K.K., Miller, T.L., and Rickert, D.A. 1994. *U.S. Geological Survey Protocol for the Collection and Processing of Surface-Water Samples for the Subsequent Determination of Inorganic Constituents in Filtered Water.* Open- File Report 94-539. Reston, VA: U.S. Geological Survey. http://pubs.usgs.gov/of/1994/0539/report.pdf. Accessed May 2013.

Horowitz, A.J., Lum, K.R., Garbarino, J.R., Hall, G.E.M., LeMieux, C., and Demas, C.R. 1996. Problems associated with using filtration to define dissolved trace element concentrations in natural water samples. *Environ. Sci. Technol.* 30:954–963.

Horsburgh, J.S., Jones, A.S., Stevens, D.K., Tarboton, D.G., and Mesner, N.O. 2010. A sensor network for high frequency estimation of water quality constituent fluxes Using surrogates. *Environ. Model. Software* 25:1031–1044.

ICMM (International Council on Mining and Metals). 2008. *Sustainable Development Framework: A Sustained Commitment to Improved Industry Performance.* London: International Council on Mining and Metals. www.icmm.com/document/429. Accessed May 2013.

Idowu, O.A. 2007. Hydrological processes of interaction between surface water and groundwater—A review. *ASSET Series B* 6:172–190.

INAP (International Network for Acid Prevention). 2009. *Global Acid Rock Drainage Guide* (GARD Guide). www.gardguide.com/. Accessed May 2013.

ITRC (Interstate Technology & Regulatory Council). 2012. Incremental Sampling Methodology. ISM-1. Washington, DC: Interstate Technology & Regulatory Council, Incremental Sampling Methodology Team. www.itrcweb.org/team/public?teamid=11. Accessed May 2013.

Jambor, J.L. 2003. Mine-waste mineralogy and mineralogical perspectives of acid-base accounting. In *Environmental Aspects of Mine Wastes.* Edited by J.L. Jambor, D.W. Blowes, and A.I.M. Ritchie. Ottawa, ON: Mineralogical Association of Canada. pp. 117–145.

Jambor, J.L., and Blowes, D.W. 1998. Theory and applications of mineralogy in environmental studies of sulfide-bearing mine wastes. In *Modern Approaches to Ore and Environmental Mineralogy*, Short Course Handbook, Vol. 27. Ottawa, ON: Mineralogical Association of Canada. pp. 367–401.

Jambor, J.L., Nordstrom, D.K., and Alpers, C.N. 2000. Metal-sulfate salts from sulfide mineral oxidation. In *Sulfate Minerals: Crystallography, Geochemistry, and Environmental Significance*, Reviews in Mineralogy and Geochemistry, Vol. 40. Edited by C.N. Alpers, J.L. Jambor, and D.K. Nordstrom. Washington, DC: Mineralogical Society of America and Geochemistry Society. pp. 303–350.

Jambor, J.L., Dutrizac, J.E., Groat, L.A., and Raudsepp, M. 2002. Static tests of neutralization potentials of silicate and aluminosilicate minerals. *Environ. Geol.* 43:1-17.

Jambor, J.L., Dutrizac, J.E., and Raudsepp, M. 2007. Measured and computed neutralization potentials from static tests of diverse rock types. *Environ. Geol.* 52:1173–1185.

Johnson, C.A., and Thornton, I. 1987. Hydrological and chemical factors controlling the concentrations of Fe, Cu, Zn, and As in a river system contaminated by acid mine drainage. *Water Resour.* 21:359–365.

Jordan, G. 2009. Sustainable mineral resources management: from regional mineral resources exploration to spatial contamination risk assessment of mining. *Environ. Geol.* 58(1):153–169.

Kearl, P.M., Korte, N.E., and Cronk, T.A. 1992. Suggested modifications to ground water sampling procedures based on observations from the colloidal borescope. *Ground Water Monit. Rev.* 12:155–160.

Kearl, P.M., Korte, N.E., Stites, M., and Baker, J. 1994. Field comparison of micropurging vs. traditional ground water sampling. Ground Water Monit. Remed. 14:183–190.

Keith, L.H. 1991. *Environmental Sampling and Analysis: A Practical Guide.* Boca Raton, FL: CRC Press, Lewis Publishers.

Keith, L.H., Brass, H.J., Sullivan, D.J., Boiani, J.A., and Alben, K.T. 2005. An introduction to the National Environmental Methods Index. *Environ. Sci. Technol.* 39:173A–176A.

Kelly, S.D., Hesterberg, D., and Ravel, B., eds. 2008. Analysis of soils and minerals using X-ray absorption spectroscopy. In *Methods of Soil Analysis*, Part 5. Mineralogical Methods. Madison, WI: Soil Science Society of America. pp. 387–464.

Kennedy, E.J. 1984. Discharge ratings at gaging stations: U.S. Geological Survey Techniques of Water-Resources Investigations, book 3. http://pubs.usgs.gov/twri/twri3-a10/. Accessed May 2013.

Kerr, P.F. 1977. *Optical Mineralogy*, 4th ed. New York: McGraw-Hill.

Kilpatrick, F.A., and Cobb, E.D. 1985. Measurement of discharge using tracers: U.S. Geological Survey Techniques of Water-Resources Investigations, book 3. http://pubs.usgs.gov/twri/twri3-a16/. Accessed May 2013.

Kim, C.S., Rytuba, J.J., and Brown, G.E., Jr. 2004. Geological and anthropogenic factors influencing mercury speciation in mine wastes: An EXAFS spectroscopy study. *Appl. Geochem.* 19:379–393.

Kimball, B.A. 1997. *Use of Tracer Injections and Synoptic Sampling to Measure Metal Loading from Acid Mine Drainage*. Fact Sheet 245-96. Reston, VA: U.S. Geological Survey. http://ut.water.usgs.gov/publications/fs245-96/). Accessed May 2013.

Kimball B.A., Runkel R.L., Walton-Day K., and Bencala K.E. 2002. Assessment of metal loads in watersheds affected by acid mine drainage by using tracer injection and synoptic sampling: Cement Creek, Colorado, USA. *Appl. Geochem.* 17:1183–1207.

Kimball, B.A., Walton-Day, K., and Runkel, R.L. 2007. Quantification of metal loading by tracer injection and synoptic sampling, 1996–2000. In *Integrated Investigations of Environmental Effects of Historical Mining in the Animas River Watershed, San Juan County, Colorado*. Edited by S.E. Church, P. von Guerard, and S.E. Finger. Professional Paper 1651. Reston, VA: U.S. Geological Survey. http://pubs.usgs.gov/pp/1651/. Accessed May 2013.

Kimball, B.A., Runkel, R.L., and Walton-Day, K. 2010. An approach to quantify sources, seasonal change, and biogeochemical processes affecting metal loading in streams: Facilitating decisions for remediation of mine drainage: *Appl. Geochem.* 25:728–740.

Klein, T.L., Thamke, J.N., Harper, D.D., Farag, A.M., and Nimick, D.A. 2003. *Water-Quality, Bed-Sediment, and Biological Data, for Streams in the Upper Prickly Pear Creek Watershed, Montana, 2001*. Open-File Report 03-032. Reston, VA: U.S. Geological Survey.

Kleinmann, R.L.P., ed. 2000. *Prediction of Water Quality at Surface Coal Mines*. Acid Drainage Technology Initiative. Morgantown, WV: The National Mine Land Reclamation Center.

Koch, G.S., Jr., and Link, R.F. 2002. *Statistical Analysis of Geological Data*. New York: Dover Publications.

Kwong, Y.T.J. 1993. *Prediction and Prevention of Acid Rock Drainage from a Geological and Mineralogical Perspective*. MEND Project 1.32.1. Ottawa, ON: Environment Canada. www.mend-nedem.org/reports/files/1.32.1.pdf. Accessed May 2013.

Kwong, Y.T.J. 2003. Comprehensive environmental ore deposit models as an aid for sustainable development. *Explor. Min. Geol.* 12:31–36.

Lapakko, K., and Berndt, M. 2003. Comparison of acid production from pyrite and jarosite. In *Proceedings, Sixth International Conference on Acid Rock Drainage (ICARD)*. Victoria, Australia: Australasian Institute of Mining and Metallurgy (AusIMM).

Lee, L., and Helsel, D. 2005. Baseline models of trace elements in major aquifers of the United States. *Appl. geochem.* 20(8):1560–1570.

LeGrand, H.E., and Rosen, L. 1998. Putting hydrogeologic site studies on track (editorial). *Ground Water* 36:193.

LeGrand, H.E., and Rosen, L. 2000. Systematic makings of early stage hydrogeologic conceptual models. *Ground Water* 38:887–893.

Lowers, H.A., Meeker, G.P., and Brownfield, I.K. 2005. *Analysis of Background Residential Dust for World Trade Center Signature Components Using Electron Microscopy and X-ray Microanalysis*. U.S. Geological Survey, Open-File Report 2005-1073. http://pubs.usgs.gov/of/2005/1073/pdf/OF2005-1073_508.pdf. Accessed May 2013.

Madrid, Y., and Zayas, Z.P. 2007. Water sampling: Traditional methods and new approaches in water sampling strategy. *Trends Anal. Chem.* 26:293–299.

Maest, A.S., Nordstrom, D.K., and LoVetere, S.H. 2004. *Questa Baseline and Pre-Mining Ground-Water Quality Investigation. 4. Historical Surface-Water Quality for the Red River Valley, New Mexico, 1965 to 2001*. Scientific Investigations Report 2004-5063. Reston, VA: U.S. Geological Survey. http://pubs.usgs.gov/sir/2004/5063/. Accessed May 2013.

Maney, J.P., and Wait, A.D. 2005. The role of representativeness in the project life cycle. *Environ. Forensics*. 6:17–19.

Manning, A.H., Verplanck, P.L., Mast, M.A., and Wanty, R.B. 2008. *Hydrogeochemical Investigation of the Standard Mine Vicinity, Upper Elk Creek Basin, Colorado*. Scientific Investigations Report 2007–5265. Reston, VA: U.S. Geological Survey. http://pubs.usgs.gov/sir/2007/5265/. Accessed May 2013.

Mason, B.J. 1992. *Preparation of Soil Sampling Protocols: Sampling Techniques and Strategies.* EPA/600/R-92/128. Washington, DC: U.S. Environmental Protection Agency, Office of Research and Development. www.epa.gov/oust/cat/mason.pdf. Accessed May 2013.

McCarthy, J., and Shevenell, L. 1998. Obtaining representative ground-water samples in a fractured and karstic formation. *Ground Water* 36:251–260.

McCarthy, J.F. and Zachara, J.M. 1989. Subsurface transport of contaminants. *Environ. Sci. Technol.* 5(23):496–502.

McDermott, H.J. 2004. *Air Monitoring for Toxic Exposures*, 2nd ed. New York: John Wiley and Sons.

McGee, J.J., and Klaus, K. 2001. Application of electron probe microanalysis to the study of geological and planetary materials. *Micros. Microanal.* 7:200–210.

McKnight, D.M., and Bencala, K.E. 1990. Chemistry of iron, aluminum, and dissolved organic material in three acidic, metal-enriched, mountain streams, as controlled by watershed and in-stream processes. *Water Resour. Res.* 26:3087–3100.

McKnight, D.M., Kimball, B.A., and Bencala, K.E. 1988. Iron photoreduction and oxidation in an acidic mountain stream. *Science* 240:637–640.

McLemore, V.T., Smith, K.S., Russell, C.C., and the Sampling and Monitoring Committee of the Acid Drainage Technology Initiative-Metal Mining Sector (ADTI-MMS). 2007. Sampling and monitoring for closure. Reviews in Engineering Geology, Vol. 17. In *Understanding and Responding to Hazardous Substances at Mine Sites in the Western United States.* Edited by J.V. DeGraff. Boulder, CO: Geological Society of America. pp. 171–180.

McLemore, V., Heizler, L., Donahue, K., and Dunbar, N. 2009a. Characterization of weathering of mine rock piles: Example from the Questa Mine, New Mexico, USA. In *Proceedings, Securing the Future and 8th International Conference on Acid Rock Drainage (ICARD).* Sweden: SveMin. http://geoinfo.nmt.edu/staff/mclemore/documents/Virginia_McLemore_P_T2_Characterization-of-weathering-of-mine-rock-piles-example-from-the-Qu.pdf. Accessed May 2013.

McLemore, V., Sweeney, D., Dunbar, N., Heizler, L. and Phillips, E. 2009b. Determining quantitative mineralogy using a combination of petrographic techniques, whole rock chemistry, and MODAN. Preprint 09-20. In *Proceedings of the SME Annual Meeting and Exhibit.* Littleton, CO: SME.

Meeker, G.P., Bern, A.M., Lowers, H.A., and Brownfield, I.K. 2005. *Determination of a Diagnostic Signature for World Trade Center Dust Using Scanning Electron Microscopy Point Counting Techniques.* U.S. Geological Survey, Open-File Report 2005-1031. http://pubs.usgs.gov/of/2005/1031/pdf/OF2005_1031_508.pdf. Accessed May 2013.

Milhous, R.T. 2007. Application of physical habitat simulation in the evaluation of physical habitat suitability. In *Integrated Investigations of Environmental Effects of Historical Mining in the Animas River Watershed, San Juan County, Colorado.* Edited by S.E. Church, P. von Guerard, and S.E. Finger. Professional Paper 1651. Reston, VA: U.S. Geological Survey. http://pubs.usgs.gov/pp/1651/. Accessed May 2013.

Miranda, M., Chambers, D., and Coumans, C. 2005. *Framework for Responsible Mining: A Guide to Evolving Standards.* Center for Science in Public Participation (CSP2). www.frameworkforresponsiblemining.org. Accessed May 2013.

Moore, O.M., and Reynolds, R.O., Jr. 1989. *X-ray Diffraction and the Identification and Analyses of Clay Minerals.* New York: Oxford University Press.

Munroe, E.A. 1999. Geology and geochemistry of waste rock piles in the Hillsboro mining district, Sierra County, New Mexico. M.S. thesis, New Mexico Institute of Mining and Technology, Socorro.

Munroe, E.A., McLemore, V.T., and Kyle, P. 1999. Waste rock pile characterization, heterogeneity, and geochemical anomalies in the Hillsboro mining district, Sierra County, New Mexico. *J. Geochem. Explor.* 67:391–405.

Munroe, E.A., McLemore, V.T., and Dunbar, N.W. 2000. Mine waste rock pile geochemistry and mineralogy in southwestern New Mexico, USA, Vol. II. In *Proceedings, 5th International Conference on Acid Rock Drainage (ICARD 2000).* Littleton, CO: SME. pp. 1327–1336.

Muste, M., Vermeyen, T.R., Hotchkiss, R., and Oberg, K.A. 2007. Acoustic velocimetry for riverine environments. *J. Hydraul. Eng.* 133:1297–1298.

Myers, J.C. 1997. *Geostatistical Error Management: Quantifying Uncertainty for Environmental Sampling and Mapping.* New York: Van Nostrand Reinhold.

Nagorski, S.A., Moore, J.N., McKinnon, T.E., and Smith, D.B. 2003. Scale-dependent temporal variations in stream water geochemistry. *Environ. Sci. Technol.* 37:859–864.

Namiesnik, J., and Szefer, P., eds. 2010. *Analytical Measurements in Aquatic Environments*. Boca Raton, FL: CRC Press.

Nesse, W.D. 2004. *Introduction to Optical Mineralogy*, 3rd ed. New York: Oxford University Press.

Neubert, J.T. 2000. *Naturally Degraded Surface Waters Associated with Hydrothermally Altered Terrane in Colorado*. Open-File Report 00–16, CD-ROM pub. Denver, CO: Colorado Geological Survey.

Neuendorf, K.K.E., Mehl, J.P., Jr., and Jackson, J.A. 2005. *Glossary of Geology*, 5th ed. Alexandria, VA: American Geological Institute.

Nielsen, D.M., ed. 2006. *Practical Handbook of Environmental Site Characterization and Ground-Water Monitoring*, 2nd ed. Boca Raton, FL: CRC Press.

Nielsen, D.M., and Nielsen, G.L. 2007. *The Essential Handbook of Ground-Water Sampling*. Boca Raton, FL: CRC Press.

Nimick, D.A., Gammons, C.H., Cleasby, T.E., Madison, J.P., Skaar, D., and Brick, C.M. 2003. Diel cycles in dissolved metal concentrations in streams: Occurrence and possible causes. *Water Resour. Res.* 39:citation1247.

Nimick, D.A., Cleasby, T.E., and McCleskey, R.B. 2005. Seasonality of diel cycles of dissolved trace-metal concentrations in a Rocky Mountain stream. *Environ. Geol.* 47:603–614.

Nimick, D.A., Gammons, C.H., and Parker, S.R. 2011. Diel biogeochemical processes and their effect on the aqueous chemistry of streams: A review. *Chem. Geol.* 283:3–17.

Niyogi, D.K., Lewis, W.M., and D.M. McKnight. 2002. Effects of stress from mine drainage on diversity, biomass, and function of primary producers in mountain streams. *Ecosystems* 5:554–567.

Nolan, K.M, Gray, J.R., and Glysson, G.D. 2005. *Introduction to Suspended-Sediment Sampling*. Scientific Investigations Report 2005-5077, CD-ROM pub. Reston, VA: U.S. Geological Survey. http://pubs.usgs.gov/sir/2005/5077/. Accessed May 2013.

Nordstrom, D.K. 2004. Modeling low-temperature geochemical processes. In *Surface and Ground Water, Weathering, and Soils: Treatise on Geochemistry*. Edited by J.I. Drever. Amsterdam: Elsevier. pp. 37–72.

Nordstrom, D.K. 2008. *Questa Baseline and Pre-Mining Ground-Water Quality Investigation. 25. Summary of Results and Baseline and Pre-Mining Ground-Water Geochemistry, Red River Valley, Taos County, New Mexico, 2001-2005*. Professional Paper 1728. Reston, VA: U.S. Geological Survey. http://pubs.usgs.gov/pp/1728/pdf/p1728_508.pdf. Accessed May 2013.

Nordstrom, D. K. 2011. Hydrogeochemical processes governing the origin, transport and fate of major and trace elements from mine wastes and mineralized rock to surface waters. *Applied Geochemistry.* 26(11):1777-1791.

Nordstrom, D.K., and Alpers, C.N. 1999. Geochemistry of acid mine waters. In *The Environmental Geochemistry of Mineral Deposits—Part A: Processes, Techniques, and Health Issues*. Reviews in Economic Geology, Vol. 6A. Littleton, CO: Society of Economic Geologists. pp. 133–160.

Nordstrom, D.K., and Munoz, J.L. 2006. *Geochemical Thermodynamics*, 2nd ed. Blackburn Press.

Nordstrom, D.K., and Wilde, F.D. 2005. Reduction-oxidation potential (electrode method). In *Field Measurements: U.S. Geological Survey Techniques of Water-Resources Investigations, book 9*. Edited by F.D. Wilde. http://water.usgs.gov/owq/FieldManual/Chapter6/6.5_v_1.2.pdf. Accessed May 2013.

Nordstrom, D.K., Nicholson, A., Verberg, R., and Mayer, U., eds. Forthcoming. *Geochemical Modeling for Mine Site Characterization and Remediation of Mine Wastes*. Management Technologies for Metal Mining Influenced Water series, Vol. 4. Englewood, CO: SME.

NPS (National Park Service). 2001. Appendix C, STORET water quality control/edit checking. In *Baseline Water Quality Data Inventory and Analysis—Rocky Mountain National Park*. Technical Report NPS/NRWRD/NRTR-2000/268. Washington, DC: National Park Service.

Nuffer, L.L., Hassig, N.L., Sego, L.H., Pulsipher, B.A., Wilson, J.E., and Matzke, B. 2009. *Validation of Statistical Sampling Algorithms in Visual Sample Plan (VSP): Summary Report*. Report PNNL-18253. Pacific Northwest National Laboratory. www.pnl.gov/main/publications/external/technical_reports/PNNL-18253.pdf. Accessed May 2013.

Nuffield, E.W. 1966. *X-ray Diffraction Methods*. New York: Wiley.

Oberg, K.A., and Mueller, D.S. 2007. Validation of streamflow measurements made with Acoustic Doppler Current Profilers. *J. Hydraul. Eng.* 133:1421–1432.

Oerter, E., Brimhall, G.H., Jr., Redmond, J., and Walker, B. 2007. A method for quantitative pyrite abundance in mine rock piles by powder X-ray diffraction and Rietveld method. *Appl. Geochem.* 22:2907–2925.

Paktunc, A.D. 1998. MODAN: An interactive computer program for estimating mineral quantities based on bulk composition. *Computers Geosci.* 24:425–431.

Paktunc, A.D. 2001. MODAN—A Computer program for estimating mineral quantities based on bulk composition: Windows version. *Computers Geosci.* 27:883–886.

Parbhakar, A., Edraki, M., Bradshaw, D., and Walters, S. 2009. Mineralogical characterization techniques for predicting acid rock drainage. In *Proceedings, Securing the Future and 8th International Conference on Acid Rock Drainage (ICARD).* Sweden: SveMin.

Parker, L.V. 1994. The effects of ground water sampling devices on water quality: A literature review. *Ground Water Monit. Remed.* 14:130–141.

Piatak, N.M., Seal II, R.R., and Hammarstrom, J.M. 2004. Mineralogical and geochemical controls on the release of trace elements from slag produced by base-and precious-metal smelting at abandoned mine sites. *Appl. Geochem.* 19(7):1039–1064.

Pirrie, D., and Rollinson, G.K. 2011. Unlocking the applications of automated mineral analysis. *Geology Today*. 27:226-235.

Pirrie, D., Butcher, A.R., Power, M.R., Gottlieb, P., and Miller, G.L. 2004. Rapid quantitative mineral and phase analysis using automated scanning electron microscopy (QEMSCAN); Potential applications in forensic geoscience. In *Forensic Geoscience; Principles, Techniques and Applications*, Geological Society Special Publication 232. London: The Geological Society. pp. 123–136.

Pirrie, D., Power, M.R., Rollinson, G.K., Wiltshire, P.E.J., Newberry, J., Campbell, H.E. 2009. Automated SEM-EDS (QEMSCAN®) mineral analysis in forensic soil investigations: Testing instrumental reproducibility. In *Criminal and Environmental Soil Forensics*. London: Springer Science. pp. 411–430.

Pitard, F.F. 1993. *Pierre Gy's Sampling Theory and Sampling Practice: Heterogeneity, Sampling Correctness, and Statistical Process Control*, 2nd ed. Boca Raton, FL: CRC Press.

Pitard, F.F. 2010. Theoretical, practical, and economic difficulties in sampling for trace constituents. *J. S. Afr. Inst. Min. Metall.* 110:313–321.

Plumlee, G. S. 1999. The environmental geology of mineral deposits. In *The Environmental Geochemistry of Mineral Deposits—Part A: Processes, Techniques, and Health Issues*. Reviews in Economic Geology, Vol. 6A. Littleton, CO: Society of Economic Geologists. pp. 71–116.

Plumlee, G. 2009. Report from ground zero—How geoscientists aid in the aftermath of environmental disasters. *Earth Mag.* October. American Geological Institute. www.earthmagazine.org/earth/article/278-7d9-a-1. Accessed May 2013.

Plumlee, G.S. and Nash, J.T. 1995. Geoenvironmental models of mineral deposits—Fundamentals and applications. In *Preliminary Compilation of Descriptive Geoenvironmental Mineral Deposit Models*. Open-File Report 95-0831. Reston, VA: U.S. Geological Survey. http://pubs.usgs.gov/of/1995/ofr-95-0831/. Accessed May 2013.

Plumlee, G.S., Smith, K.S., and Ficklin, W.H. 1994. *Geoenvironmental Models of Mineral Deposits, and Geology-Based Mineral-Environmental Assessments of Public Lands*. Open-File Report 94-203. Reston, VA: U.S. Geological Survey. http://pubs.usgs.gov/of/1994/ofr-94-0203/94-203.pdf. Accessed May 2011.

Plumlee, G.S., Streufert, R.K., Smith, K.S., Smith, S.M., Wallace, A.R., Toth, M., Nash, J.T., Robinson, R.A., Ficklin, W.H., and Lee, G.K. 1995. *Map Showing Potential Metal-Mine Drainage Hazards in Colorado, Based on Mineral-Deposit Geology* (1 over-size sheet, scale 1:750,000). Open-File Report 95-26. Reston, VA: U.S. Geological Survey. Map Series 29. Denver, CO: Colorado Geological Survey. http://geosurveystore.state.co.us/p-986-potential-metal-mine-drainage-hazards-in-co-based-on-mineral-deposit-geology.aspx. Accessed May 2013.

Plumlee G.S., Smith, K.S., Montour, M.R., Ficklin, W.H., and Mosier, E.L. 1999. Geologic controls on the composition of natural waters and mine waters draining diverse mineral-deposit types. In *The Environmental Geochemistry of Mineral Deposits—Part B: Case Studies and Research Topics*. Reviews in Economic Geology, Vol. 6B. Littleton, CO: Society of Economic Geologists. pp. 373–432.

Power, M. 2002. Assessing fish population response to stress. In *Biological Indicators of Aquatic Ecosystem Stress*. Edited by S.M. Adams. Bethesda, MD: American Fisheries Society.

Price, W.A. 2009. *Prediction Manual for Drainage Chemistry from Sulphidic Geologic Materials*. MEND Report 1.20.1. Ottawa, ON: Natural Resources C www.mend-nedem.org/reports/files/1.20.1.pdf. Accessed May 2013.

Puls, R.W., and M.J. Barcelona. 1989. Filtration of ground water samples for metals analyses. *Haz. Waste Haz. Mater.* 6(4):385–393.

Puls, R.W., and Barcelona, M.J. 1996. *Low-Flow (Minimal Drawdown) Ground-Water Sampling Procedures.* Ground Water Issue. EPA/540/S-95/504. Washington, DC: U.S. Environmental Protection Agency. www.epa.gov/superfund/remedytech/tsp/download/lwflw2a.pdf. Accessed May 2013.

Puls, R.W., and Powell, R.M. 1992. Acquisition of representative ground water quality samples for metals. *Ground Water Monit. Rev.* 12:167–176.

Puls, R.W., J.H. Eychaner, and R.M. Powell. 1990. *Colloidal-Facilitated Transport of Inorganic Contaminants in Ground Water: Part I. Sampling Considerations.* EPA/600/M-90/023. NTIS PB 91-168419. Washington, DC: U.S. Environmental Protection Agency.

Puls, R.W., D.A. Clark, B. Bledsoe, R.M. Powell, and C.J. Paul. 1992. Metals in Ground Water: Sampling Artifacts and Reproducibility. *Haz. Waste Haz. Mater.* 9(2):149–162.

Rainbow, P.S. 2007. Trace metal bioaccumulation: Models, metabolic availability and toxicity. *Environ. Int.* 33:576–582.

Ramsey, C.A., and Hewitt, A.D. 2005. A methodology for assessing sample representativeness. *Environ. Forensics* 6:71–75.

Ramsey, C.A., and Suggs, J. 2001. Improving laboratory performance through scientific subsampling techniques. *Environ. Test. Anal.* 10:12–16.

Rantz, S.E., and others. 1982. *Measurement and Computation of Streamflow: Volume 1. Measurement of Stage and Discharge. Volume 2. Computation of Discharge.* Water-Supply Paper 2175. Reston, VA: U.S. Geological Survey. http://pubs.usgs.gov/wsp/wsp2175/. Accessed May 2013.

Ranville, J.F. and Schmiermund, R.L. 1999. General aspects of aquatic colloids in environmental geochemistry. In *The Environmental Geochemistry of Mineral Deposits—Part A: Processes, Techniques, and Health Issues. Reviews in Economic Geology,* Vol. 6A. Littleton, CO: Society of Economic Geologists. pp. 183–199.

Ranville, J., Blumenstein, E., Adams, M., Choate, L., Smith, K., and Wildeman, T. 2006. Integrating bioavailability approaches into waste rock evaluations. In *Proceedings, 7th International Conference on Acid Rock Drainage (ICARD 7).* Lexington, KY: American Society of Mining and Reclamation. pp. 1642–1653. www.imwa.info/docs/imwa_2006/1642-Ranville-CO.pdf. Accessed May 2013.

Reed, S.J.B. 1996. *Electron Microprobe Analysis and Scanning Electron Microscopy in Geology.* New York: Cambridge University Press.

Reimann, C., and Garrett, R.G. 2005. Geochemical background—concept and reality. *Scie. Total Environ.* 350(1):12–27.

Reynolds, J.B. 1996. Electrofishing. In *Fisheries Techniques,* 2nd ed. Edited by B.R. Murphy and D.W. Willis. Bethesda, MD: American Fisheries Society.

Rice, E.W., Baird, R.B., Eaton, A.D., and Clesceri, L.S., eds. 2012. *Standard Methods for the Examination of Water and Wastewater,* 22nd edition. Washington, D.C.: American Public Health Association, American Water Works Association, and Water Environment Federation.

Ritchey, J. 2002. Low-flow purging and sampling ground water, evolution of technology and standards. *ASTM Stand. News,* April 2002, pp. 18–23.

Riviere, J.C., and Myhra, S., eds. 2009. *Handbook of Surface and Interface Analysis—Methods for Problem-Solving,* 2nd ed. Boca Raton, FL: CRC Press.

Rockwell, B.W. 2011. *Description and Validation of an Automated Methodology for Mapping Mineralogy, Vegetation, and Hydrothermal Alteration Type from ASTER Satellite Imagery with Examples from the San Juan Mountains, Colorado.* Scientific Investigations Map 3190. Reston, VA: U.S. Geological Survey. http://pubs.usgs.gov/sim/3190/. Accessed May 2013.

Rockwell, B.W. 2013. Comparative mineral mapping in the Colorado Mineral Belt using AVIRIS and ASTER remote sensing data: U.S. Geological Survey Scientific Investigations Map 3256. pamphlet, 1 map sheet, scale 1:150,000, http://pubs.usgs.gov/sim/3256/. Accessed July 2013.

Roig, B. 2007. Report on laboratory and field validations of screening tools based on performance criteria evaluation. In *Screening Methods for Water Data Information in Support of the Implementation of the Water Framework Directive,* Deliverable D44.

Rollinson, H.R. 1993. *Using Geochemical Data: Evaluation, Presentation, Interpretation.* Harlow, England: Pearson Education Ltd., Prentice Hall, Longman Group.

Ruby, M.V., Davis, A., Link, T.E., Schoof, R., Chaney, R.L., Freman, G.B., and Bergstrom, P. 1993. Development of an in vitro screening test to evaluate the in vivo bioaccessibility of ingested mine-waste lead. *Environ. Sci. Technol.* 27(13):2870–2877.

Runkel, R.L., Kimball, B.A., Walton-Day, K., and Verplanck, P.L. 2007. A simulation-based approach for estimating premining water quality: Red Mountain Creek, Colorado. *Appl. Geochem.* 22(9):1899–1918.

Runnells, D.D., Shepard, T.A., Angino, E.E. 1992. Metals in water: determining natural background concentrations in mineralized areas. *Environ. Sci. Technol.* 26:2316–2322.

Runnells, D.D., Shields, M.J., and Jones, R.L. 1997. Methodology for adequacy of sampling of mill tailings and mine waste rock. In *Proceedings, Tailings and Mine Waste '97*. Rotterdam: Balkema. pp. 561–563.

Russell, C. 2006. Mine drainage challenges from A to Z in the United States. In *Proceedings, 7th International Conference on Acid Rock Drainage (ICARD 7)*. Lexington, KY: American Society of Mining and Reclamation. pp. 1785–1802. www.asmr.us/Publications/Conference%20Proceedings/2006/1785-Russell-CO.pdf. Accessed May 2013.

Santore, R.C., Di Toro, D.M., Paquin, P.R., Allen, H.E., and Meyer, J.S. 2001. Biotic ligand model of the acute toxicity of metals. 2. Application to acute copper toxicity in freshwater fish and Daphnia. *Environ. Toxicol. Chem.* 20:2397–2402.

Schmidt, T.S., Clements, W.H., Wanty, R.B., Verplanck, P.L., Church, S.E., San Juan, C.A., and Klein, T.L. 2012. Geologic processes influence the effects of mining on aquatic ecosystems. *Ecolog. Appl.* 22(3):870–879.

Schmiermund, R.L., and Drozd, M.A. 1997. Acid mine drainage and other mining influenced waters (MIW). In *Mining Environmental Handbook*. Edited by J.J. Marcus. London: Imperial College Press. pp. 599–617.

Schmiermund, R.L., and Ranville, J. 2006. Application of geo-environmental and biotic ligand models in the life cycle of mine development—A flowchart for integration. In *Proceedings, 7th International Conference on Acid Rock Drainage (ICARD 7)*. Lexington, KY: American Society of Mining and Reclamation. pp. 1920–1925. www.imwa.info/docs/imwa_2006/1920 Schmiermund-CO-3.pdf. Accessed May 2013.

Schmiermund, R.L., Lazo, M.C., and Parnow, C.C. 2006. Application of geo-environmental models to accelerated EIA and permitting processes for an Andean porphyry Cu-Au deposit. In *Proceedings, 7th International Conference on Acid Rock Drainage (ICARD 7)*. Lexington, KY: American Society of Mining and Reclamation. pp. 1903–1911. www.imwa.info/docs/imwa_2006/1903-Schmiermund-CO-1.pdf. Accessed May 2013.

Schreuder, H.T., Ernst, R., and Ramirez-Maldonado, H. 2004. *Statistical Techniques for Sampling and Monitoring Natural Resources*. General Technical Report-RMRS-GTR-126. Fort Collins, CO: U.S. Department of Agriculture, Rocky Mountain Research Station. www.fs.fed.us/rm/pubs/rmrs_gtr126.pdf. Accessed May 2013.

Schumacher, B.A., Shines, K.C., Burton, J.V., and Papp, M.L. 1990. *A Comparison of Soil Sample Homogenization Techniques*. EPA 600//X-90/043. Washington, DC: U.S. Environmental Protection Agency, Office of Research and Development. www.epa.gov/esd/cmb/research/papers/bs120.pdf. Accessed May 2013.

Scullion, J., and Edwards, R.W. 1980. The effect of pollutants from the coal industry on the fish fauna of a small river in the South Wales coalfield. *Environ. Pollut. (Series A)* 21:141–153.

Seal, R.R. II, and Foley, N.K., eds. 2002. *Progress on Geoenvironmental Models for Selected Mineral Deposit Types*. Open-File Report 02-195. Reston, VA: U.S. Geological Survey. http://pubs.usgs.gov/of/2002/of02-195/. Accessed May 2013.

Seal, R.R. II, and Hammarstrom, J.M. 2003. Geoenvironmental models of mineral deposits: Examples from massive sulfide and gold deposits. In *Environmental Aspects of Mine Wastes*. Edited by J.L. Jambor, D.W. Blowes, and A.I.M. Ritchie. Short Course Series, Vol. 31. Ottawa, ON: Mineralogical Association of Canada.

Seal, R.R., II, Piatak, N.M., Levitan, D.M., Hageman, P.L., and Hammarstrom, J.M. 2009. Comparison of geochemical characteristics of modern-style mine wastes from a variety of mineral deposit types for insights into environmental challenges associated with future mining. In *Proceedings, Securing the Future and 8th International Conference on Acid Rock Drainage (ICARD)*. Sweden: SveMin.

Sengupta, M. 1990. *Mine Environmental Engineering, Vol. I*. Boca Raton, FL: CRC Press.

Sevee, J.E., White, C.A., and Maher, D.J. 2000. An analysis of low-flow ground water sampling methodology. *Ground Water Monit. Remed.* 20(2):87–93.

Shaw, S., and Mills, C. 2012. Petrology and mineralogy in ARD prediction. Infomine. http://technology.infomine.com/enviromine/ard/mineralogy/petrology%20and%20mineralogy.htm. Accessed May 2013.

Skousen J.G., J.C. Sencindiver, and R.M. Smith. 1987. *A Review of Procedures for Surface Mining and Reclamation in Areas with Acid-Producing Materials, in cooperation with the West Virginia Surface Mine Drainage Task Force, the West Virginia University Energy and Water Research Center, and the West Virginia Mining and Reclamation Association.* Morgantown, WV: The Center.

Sliwka-Kaszyńska, M., Kot-Wasik, A., and Namieśnik, J. 2003. Preservation and storage of water samples. *Crit. Rev. Environ. Sci. Technol.* 33:31–44.

Smith, K.S. 1999. Metal sorption on mineral surfaces: An overview with examples relating to mineral deposits. In *The Environmental Geochemistry of Mineral Deposits—Part A: Processes, Techniques, and Health Issues*. Reviews in Economic Geology, Vol. 6A. Littleton, CO: Society of Economic Geologists. pp. 161–182.

Smith, K.S. 2007. Strategies to predict metal mobility in surficial mining environments. Reviews in Engineering Geology, Vol. 17. In *Understanding and Responding to Hazardous Substances at Mine Sites in the Western United States*. Edited by J.V. DeGraff. Boulder, CO: Geological Society of America. pp. 25–45.

Smith, K.S. 2011. Approach for environmental baseline water sampling. In *Proceedings of the 2011 SME Annual Meeting and Exhibit*. Preprint 11-157. Littleton, CO: SME.

Smith, K.S., and Huyck, H.L.O. 1999. An overview of the abundance, relative mobility, bioavailability, and human toxicity of metals. In *The Environmental Geochemistry of Mineral Deposits—Part A: Processes, Techniques, and Health Issues*. Reviews in Economic Geology, Vol. 6A. Littleton, CO: Society of Economic Geologists. pp. 29–70.

Smith, K.S., Ramsey, C.A., and Hageman, P.L. 2000. Sampling strategy for the rapid screening of mine-waste dumps on abandoned mine lands. In *Proceedings, 5th International Conference on Acid Rock Drainage (ICARD 2000)*. Littleton, CO: SME. pp. 1453–1461. http://crustal.usgs.gov/projects/minewaste/pdfs/kathy.pdf. Accessed May 2013.

Smith, K.S., Campbell, D.L., Desborough, G.A., Hageman, P.L., Leinz, R.W., Stanton, M.R., Sutley, S.J., Swayze, G.A., and Yager, D.B. 2002. Toolkit for the rapid screening and characterization of waste piles on abandoned mine lands. In *Progress on Geoenvironmental Models for Selected Mineral Deposit Types*. Edited by R.R. Seal II, and N.K. Foley. Open-File Report 02-0195. Reston, VA: U.S. Geological Survey. pp. 55–64. http://pubs.usgs.gov/of/2002/of02-195/. Accessed May 2013.

Smith, K.S., Ranville, J.F., Adams, M.K., Choate, L.M., Church, S.E., Fey, D.L., Wanty, R.B., and Crock, J.G. 2006. Predicting toxic effects of copper on aquatic biota in mineralized areas by using the biotic ligand model. In *Proceedings, 7th International Conference on Acid Rock Drainage (ICARD 7)*. Lexington, KY: American Society of Mining and Reclamation. pp. 2055–2077. www.imwa.info/docs/imwa_2006/2055-Smith-CO.pdf. Accessed May 2013.

Smith, K.S., Hageman, P.L., Briggs, P.H., Sutley, S.J., McCleskey, R.B., Livo, K.E., Verplanck, P.L., Adams, M.G., and Gemery-Hill, P.A. 2007. *Questa Baseline and Pre-Mining Ground-Water Quality Investigation: 19. Leaching Characteristics of Composited Materials from Mine Waste-Rock Piles and Naturally Altered Areas near Questa, New Mexico*. Scientific Investigations Report 2006-5165. Reston, VA: U.S. Geological Survey. http://pubs.usgs.gov/sir/2006/5165/. Accessed May 2013.

Smith, K.S., Ranville, J.F., Diedrich, D.J., McKnight, D.M., and Sofield, R.M. 2009. Consideration of iron-organic matter interactions when predicting aquatic toxicity of copper in mineralized areas. In *Proceedings, Securing the Future and 8th International Conference on Acid Rock Drainage (ICARD)*. Sweden: SveMin. www.proceedings-stfandicard-2009.com/pdfer/Kathleen_Smith_B5_T6_Consideration-of-Iron-Organic-Matter-Interactions-when-Predicting-Aquatic-Toxicity-of-Copper-in-Mineralized-Areas.pdf. Accessed May 2013.

Smith, K.S., Hoal, K.O., Walton-Day, K., Stammer, J.G., and Pietersen, K. 2013. Automated quantitative micro-mineralogical characterization for environmental applications. In *Proceedings, 2013 SME Annual Meeting and Exhibit*. Paper 13-140. Englewood, CO: SME.

Sobek, A.A., Schuller, W.A., Freeman, J.R., and Smith, R.M. 1978. *Field and Laboratory Methods Applicable to Overburden and Minesoils*. U.S. Environmental Protection Agency, Office of Research and Development. EPA-600/2-78-054.

Sparks, D.L. 2003. *Environmental Soil Chemistry*, 2nd ed. Boston: Academic Press.

Sparks, D.L. 2005. Toxic metals in the environment: The role of surfaces. *Elements* 1:193–197.

Stoddard, J.L., Larsen, D.P., Hawkins, C.P., Johnson, R.K., and Norris, R.H. 2006. Setting expectations for the ecological condition of streams: The concept of reference condition. *Ecol. Appl.* 16:1267–1276.

Stribling, J.B., Jessup, B.K., and Feldman, D.L. 2008. Precision of benthic macroinvertebrate indicators of stream condition in Montana. *J. N. Am. Benthol. Soc.* 27(1):58–67.

Sullivan, A.B., and Drever, J.I. 2001. Spatiotemporal variability in stream chemistry in a high-elevation catchment affected by mine drainage. *J. Hydrol.* 252:240–253.

Swayze, G.A., Smith, K.S., Clark, R.N., Sutley, S.J., Pearson, R.M., Vance, J.S., Hageman, P.L., Briggs, P.H., Meier, A.L., Singleton, M.J., and Roth, S. 2000. Using imaging spectroscopy to map acidic mine waste. *Environ. Sci. Technol.* 34:47–54.

Taylor, J.K. 1987. *Quality Assurance of Chemical Measurements*. Chelsea, MI: Lewis Publishers, CRC Press.

Tessier, A., Campbell, P.G.C., and Bisson, M. 1979. Sequential extraction procedure for the speciation of particulate trace metals. *Anal. Chem.* 51:844–851.

Thompson, M. 1989. Analytical quality control in theory and practice. In *Proceedings, 3rd International Symposium on the Harmonization of Quality Assurance Systems in Analytical Chemistry*. Washington, DC: ISO/REMCO 184. pp. 183–189.

Todd, A.S., McKnight, D.M., Jaros, C.L., and Marchitto, T.M. 2007. Effects of acid rock drainage on stocked rainbow trout (O. mykiss): An in-situ, caged fish experiment. *Environ. Monit. Assess.* 130:111–127.

Toth, J. 1963. A theoretical analysis of groundwater flow in small drainage basins. *J.Geophys. Res.* 68(16):4795–4812.

Tremblay, G.A., and Hogan, C.M., eds. 2001a. *MEND Manual, Vol. 2—Sampling and Analysis*. MEND document 5.4.2b. Ottawa, ON: Natural Resources Canada, Mine Environment Neutral Drainage Program.

Tremblay, G.A., and Hogan, C.M., eds. 2001b. *MEND Manual, Vol. 6—Monitoring*. MEND document 5.4.2f. Ottawa, ON: Natural Resources Canada, Mine Environment Neutral Drainage Program.

Triad Resource Center. n.d. Categorizing analytical technologies. www.triadcentral.org/mgmt/meas/key/tech/index.cfm. Accessed 10/4/13.

Truckenbrodt, D., and Einax, J. 1995. Sampling representativity and homogeneity of river sediments. *Fresenius J. Anal. Chem.* 352:437–443.

Turnipseed, D.P., and Sauer, V.B. 2010. *Discharge Measurements at Gaging Stations*. Techniques and Methods, book 3. Reston, VA: U.S. Geological Survey. http://pubs.usgs.gov/tm/tm3-a8/. Accessed May 2013.

USEPA (U.S. Environmental Protection Agency). 1983. *Methods for Chemical Analysis of Water and Wastes*: U.S. Environmental Protection Agency, EPA 600479020. http://yosemite.epa.gov/water/owrccatalog.nsf/1ffc8769fdecb48085256ad3006f39fa/f995d73aff86600085256b0600724015!OpenDocument. Accessed May 2013.

USEPA (U.S. Environmental Protection Agency). 1984. *Characterization of Hazardous Waste Sites–A Methods Manual: Vol. II, Available Sampling Methods*, 2nd ed.: U.S. Environmental Protection Agency, EPA-600/4-84-076.

USEPA (U.S. Environmental Protection Agency). 1991. *Handbook: Ground Water Volume II: Methodology*: U.S. Environmental Protection Agency, EPA/625/6-90/016b.

USEPA (U.S. Environmental Protection Agency). 1995. *Superfund Program Representative Sampling Guidance–Volume 1: Soil*: U.S. Environmental Protection Agency, EPA 540/R-95/141 PB96-963207. www.clu-in.org/download/char/SF_Rep_Samp_Guid_soil.pdf. Accessed May 2013.

USEPA (U.S. Environmental Protection Agency). 1997. Memorandum (November 5, 1997): *Establishing Site Specific Aquatic Life Criteria to Natural Background*. U.S. Environmental Protection Agency, Tudor T. Davies, Director, Office of Science and Technology. http://water.epa.gov/scitech/swguidance/standards/upload/2009_01_29_criteria_naturalback.pdf. Accessed May 2013.

USEPA (U.S. Environmental Protection Agency). 1999. *Correct Sampling Using the Theories of Pierre Gy*. Technology Support Center Fact Sheet 197CMB98.FS-14 (Rev. 4/26/99): U.S. Environmental Protection Agency, Environmental Sciences Division. www.epa.gov/esd/factsheets/csutpg.pdf. Accessed May 2013.

USEPA (U.S. Environmental Protection Agency). 2000. *Abandoned Mine Site Characterization and Cleanup Handbook*: U.S. Environmental Protection Agency, EPA 910-B-00-001. www.epa.gov/superfund/policy/remedy/pdfs/amscch. Accessed December 2013.

USEPA (U.S. Environmental Protection Agency). 2001a. *Data Quality Objectives Decision Error Feasibility Trials Software (DEFT)–User's Guide* (EPA QA/G-4D): U.S. Environmental Protection Agency, EPA/240/B-01/007. www.epa.gov/quality/qs-docs/g4d-final.pdf. Accessed May 2013.

USEPA (U.S. Environmental Protection Agency). 2001b. *EPA Requirements for Quality Assurance Project Plans* (EPA QA/R-5): U.S. Environmental Protection Agency, EPA/240/B-01/003. www.epa.gov/quality/qs-docs/r5-final.pdf. Accessed May 2013.

USEPA (U.S. Environmental Protection Agency). 2001c. *Performing Quality Flow Measurements at Mine Sites*: U.S. Environmental Protection Agency, Office of Research and Development, EPA/600/R-01/043. http://nepis.epa.gov/Adobe/PDF/30002H0Y.pdf. Accessed May 2013.

USEPA (U.S. Environmental Protection Agency). 2001d. *Abandoned Mine Site Characterization and Cleanup Handbook*: U.S. Environmental Protection Agency, EPA 530-C-01-001.

USEPA (U.S. Environmental Protection Agency). 2002a. *Guidance on Choosing a Sampling Design for Environmental Data Collection for Use in Developing a Quality Assurance Project Plan* (EPA QA/G-5S): U.S. Environmental Protection Agency, EPA/240/R-02/005. www.epa.gov/quality/qs-docs/g5s-final.pdf. Accessed May 2013.

USEPA (U.S. Environmental Protection Agency). 2002b. *Methods for Measuring the Acute Toxicity of Effluents and Receiving Waters to Freshwater and Marine Organisms*, Fifth Edition: U.S. Environmental Protection Agency, Office of Water, EPA-821-R-02-012. http://water.epa.gov/scitech/methods/cwa/wet/upload/2007_07_10_methods_wet_disk2_atx.pdf. Accessed May 2013.

USEPA (U.S. Environmental Protection Agency). 2003a. *Diesel Exhaust in the United States*: U.S. Environmental Protection Agency, EPA420-F-03-022. www.epa.gov/cleandiesel/documents/420f03022.pdf. Accessed May 2013.

USEPA (U.S. Environmental Protection Agency). 2003b. *EPA and Hardrock Mining: A Source Book for Industry in the Northwest and Alaska*: U.S. Environmental Protection Agency, Region 10. http://yosemite.epa.gov/R10/WATER.NSF/840a5de5d0a8d1418825650f00715a27/e4ba15715e97ef2188256d2c00783a8e/$FILE/Maintext.pdf. Accessed May 2013.

USEPA (U.S. Environmental Protection Agency). 2005. *Procedures for the Derivation of Equilibrium Partitioning Sediment Benchmarks (ESBs) for the Protection of Benthic Organisms: Metal Mixtures (Cadmium, Copper, Lead, Nickel, Silver and Zinc)*: U.S. Environmental Protection Agency, EPA-600-R-02-011. www.epa.gov/nheerl/download_files/publications/metalsESB_022405.pdf. Accessed May 2013.

USEPA (U.S. Environmental Protection Agency). 2006a. *Data Quality Assessment: Statistical Methods for Practitioners* (EPA QA/G-9S): U.S. Environmental Protection Agency, EPA/240/B-06/003. www.epa.gov/quality/qs-docs/g9s-final.pdf. Accessed May 2013.

USEPA (U.S. Environmental Protection Agency). 2006b. *Guidance on Systematic Planning Using the Data Quality Objectives Process* (EPA QA/G-4): U.S. Environmental Protection Agency, EPA/240/B-06/001. www.epa.gov/quality/qs-docs/g4-final.pdf. Accessed May 2013.

USEPA (U.S. Environmental Protection Agency). 2006c. *Wadeable Streams Assessment: A Collaborative Survey of the Nation's Streams*: U.S. Environmental Protection Agency, EPA 841-B-06-002. http://water.epa.gov/type/rsl/monitoring/streamsurvey/upload/2007_5_16_streamsurvey_WSA_Assessment_May2007-2.pdf. Accessed May 2013.

USEPA (U.S. Environmental Protection Agency). 2007a. *Aquatic Life Ambient Freshwater Quality Criteria–Copper (2007 Revision)*: U.S. Environmental Protection Agency, EPA-822-R-07-001. http://water.epa.gov/scitech/swguidance/standards/criteria/aqlife/pollutants/copper/upload/2009_04_27_criteria_copper_2007_criteria-full.pdf. Accessed May 2013.

USEPA (U.S. Environmental Protection Agency). 2007b. *Framework for Metals Risk Assessment*: U.S. Environmental Protection Agency, EPA 120/R-07/001. www.epa.gov/raf/metalsframework/pdfs/metals-risk-assessment-final.pdf. Accessed May 2013.

USEPA (U.S. Environmental Protection Agency). 2007c. *Guidance for Preparing Standard Operating Procedures (SOPs)* (EPA QA/G-6): U.S. Environmental Protection Agency, EPA/600/B-07/001. www.epa.gov/quality/qs-docs/g6-final.pdf. Accessed May 2013.

USEPA (U.S. Environmental Protection Agency). 2007d. *Remediation System Evaluation (RSE)–Central City/Clear Creek Superfund Site, Argo Tunnel Water Treatment Plant, Idaho Springs, Colorado*: U.S. Environmental Protection Agency, EPA 542-R-07-019. www.epa.gov/tio/download/remed/rse/final_argo_wtp_rse.pdf. Accessed May 2013.

USEPA (U.S. Environmental Protection Agency). 2007e. Table 3-2, revision 4. *Test Methods for Evaluating Solid Waste, Physical/Chemical Methods (SW-846)*: U.S. Environmental Protection Agency. www.epa.gov/wastes/hazard/testmethods/sw846/. Accessed May 2013.

USEPA (U.S. Environmental Protection Agency). 2008. *Final Remedial Investigation Report for the Gilt Edge Superfund Site, Lawrence County, South Dakota*: U.S. Environmental Protection Agency, U.S. EPA Contract No. 68-W-05-049. www.epa.gov/region8/superfund/sd/giltedge/RI_ReportGiltEdgeVolumeI_Text.pdf. Accessed May 2013.

USEPA (U.S. Environmental Protection Agency). 2012. Quality assurance glossary. www.epa.gov/nrmrl/qa/glossary.html. Accessed 10/4/13.

USEPA (U.S. Environmental Protection Agency). Variously dated. *Test Methods for Evaluating Solid Waste, Physical/Chemical Methods (SW-846)*: U.S. Environmental Protection Agency. www.epa.gov/wastes/hazard/testmethods/sw846/. Accessed May 2013.

USGS (U.S. Geological Survey). 2011. Chemical modeling of acid waters: Questa baseline and pre-mining ground-water quality investigation, Red River Valley Basin, New Mexico. Home page. wwwbrr.cr.usgs.gov/projects/GWC_chemtherm/questa.htm. Accessed 9/30/13.

USGS (U.S. Geological Survey). 2013a. Laser ablation ICP-MS laboratory. http://crustal.usgs.gov/laboratories/icpms/laser_ablation.html. Accessed 12/6/13..

USGS (U.S. Geological Survey). 2013b. Upper Animas River study area reports, abstracts, and papers. http://amli.usgs.gov/reports/animas_rpts.html, Accessed 9/30/13..

USGS (U.S. Geological Survey). Variously dated. *National Field Manual for the Collection of Water-Quality Data*. Techniques of Water-Resources Investigations, book 9, chaps. A1-A9. Reston, VA: U.S. Geological Survey. http://water.usgs.gov/owq/FieldManual/. Accessed May 2013.

van der Plas, L., and Tobi, A.C. 1965. A chart for judging the reliability of point counting results. *Am. J. Sci.* 263:87–90.

Vaughan, D.J., and Wogelius, R.A., eds. 2000. *Environmental Mineralogy*. Notes in Mineralogy, Vol. 2. European Mineralogical Union. Budapest: Eötvös University Press.

Verplanck, P.L., Nordstrom, D.K., Bove, D.J., Plumlee, G.S., and Runkel, R.L. 2009. Naturally acidic surface and ground waters draining porphyry-related mineralized areas of the Southern Rocky Mountains, Colorado and New Mexico. *Appl. Geochem.* 24(2):255–267.

Voss, C., and Letient, H. 2006. Peruvian mine operation using dynamic system modeling. *Southwest Hydrol.* July/August 22–23, pp. 31.

Vrana, B., Mills, G.A., Allan, I.J., Dominiak, E., Svensson, K., Knutsson, J., Morrison, G., and Greenwood, R. 2005. Passive sampling techniques for monitoring pollutants in water. *Trends Anal. Chem.* 24:845–868.

Wagner, R.E. 1998. *Guide to Environmental Analytical Methods*. New York: Genium Publishing.

Wait, D., and Ramsey, C. 2007. The measurement process. In *Introduction to Environmental Forensics*, 2nd ed. Edited by B.L. Murphy and R.D. Morrison. Amsterdam; Boston: Elsevier Academic Press. pp. 83–128.

Walker, S.W., Jamieson, H.E., Lanzirotti, A., and Andrade, C.F. 2005. Determining arsenic speciation in iron oxides derived from a gold-roasting operation: Application of synchrotron micro-XRD and micro-XANES at the grain scale. *Can. Mineral.* 43:1205–1224.

Walsh, M.E., Ramsey, C.A., and Jenkins, T.F. 2002. The effect of particle size reduction by grinding on subsampling variance for explosives residues in soil. *Chemosphere* 49:1267–1273.

Walsh, M.R. 2009. *User's Manual for the CRREL Multi-Increment Sampling Tool. ERDC/CRRELL SR-09-1*. Hanover, NH: U.S. Army Engineer Research and Development Center (ERDC), Cold Regional Research and Engineering Laboratory (CRREL). www.itrcweb.org/ism-1/references/umcrrel.pdf. Accessed May 2013.

Walton-Day, K., and Poeter, E. 2009. Investigating hydraulic connections and the origin of water in a mine tunnel using stable isotopes and hydrographs. *Appl. Geochem.* 24(12):2266–2282.

Walton-Day, K., Runkel, R.L., and Kimball, B.A. 2012. Using spatially detailed water-quality data and solute-transport modeling to support total maximum daily load development. *J Am Water Resourc. Assn.* 48(5):949–969.

Wanekaya, A.K., Chen, W., and Mulchandani, A. 2008. Recent biosensing developments in environmental security. *J. Environ. Monit.* 10:703–712.

Wanty, R.B., Berger, B.R., and Plumlee, G.S. 1999. Environmental models of mineral deposits—A state of the art. *Geologica Hungarica Series Geologica*. 24:97–106. http://crustal.usgs.gov/projects/gem/pdfs/wanty2.pdf. Accessed May 2013.

Wanty, R.B., Goldhaber, M.B., Morrison, J.M., and Lee, L. 2009a. Regional variations in water quality and relationships to soil and bedrock weathering in the southern Sacramento Valley, California, USA. *Appl. Geochem*. 24:1512–1523.

Wanty, R.B., Verplanck, P.L., San Juan, C.A., Church, S.E., Schmidt, T.S., Fey, D.L., DeWitt, E.H., and Klein, T.L. 2009b. Geochemistry of surface water in alpine catchments in central Colorado, USA: Resolving host-rock effects at different spatial scales. *Appl. Geochem*. 24:600–610.

Ward, C.R., Taylor, J.C., and Cohen, D.R. 1999. Quantitative mineralogy of sandstones by X-ray diffractometry and normative analysis. *J. Sed. Res*. 69(5):1050–1062.

White, J.C., ed. 1985. *Applications of Electron Microscopy in the Earth Sciences*. Short Course Vol. 11. Ottawa, ON: Mineralogical Association of Canada.

Wildeman, T.R., Smith, K.S., and Ranville, J.F. 2007. A simple scheme to determine potential aquatic metal toxicity from mining wastes. *Environ. Forensics J*. 8:119–128.

Wilkin, R.T., McNeil, M.S., Adair, C.J., and Wilson, J.T. 2001. Field measurement of dissolved oxygen: A comparison of methods. *Ground Water Monit. Remed*. 21(4):124–132.

Williams, R.D., and Diehl, S.F., eds. 2014. *Techniques for Predicting Metal Mining Influenced Water*. Management Technologies for Metal Mining Influenced Water series, Vol. 5. Englewood, CO: SME.

Winter, T.C. 1999. Relation of streams, lakes, and wetlands to groundwater flow systems. *Hydrogeol. J*. 7:28–45.

Winter, T.C., Harvey, J.W., Franke, O.L., and Alley, W.M. 1998. *Ground Water and Surface Water—A Single Resource*. Circular 1139. Reston, VA: U.S. Geological Survey. http://water.usgs.gov/pubs/circ/circ1139. Accessed May 2013.

Winterbourn, M.J., McDiffett, W.R, and Eppley, S.J. 2000. Aluminium and iron burdens of aquatic biota in New Zealand streams contaminated by acid mine drainage: Effects of trophic level. *Sci. Total Environ*. 254:45–54.

Wolf, R.E. 2005. What is ICP-MS? ...and more importantly, what can it do? U.S. Geological Survey. http://crustal.usgs.gov/laboratories/icpms/intro.html. Accessed 12/6/13.

Wolkersdorfer, C. 2002. Mine water tracing. In *Mine Water Hydrogeology and Geochemistry*. Edited by P.L. Younger and N.S. Robins. Special Pub. 198. London: The Geological Society. pp. 47–60.

Yager, D.B., Johnson, R.H., Rockwell, B.W., Caine, J.S., and Smith, K.S. 2013. A GIS and statistical approach to identify variables that control water quality in hydrothermally altered and mineralized watersheds, Silverton, Colorado, USA. *Environ. Earth Sci*. 70(3):1057–1082. http://link.springer.com/article/10.1007/s12665-013-2229-y. Accessed May 2013. DOI:10.1007/s12665-013-2229-y

Yeskis, D., and Zavala, B. 2002. *Ground-Water Sampling Guidelines for Superfund and RCRA Project Managers*. EPA/542/S-02/001. Ground Water Forum Issue Paper. Washington, DC: U.S. Environmental Protection Agency. www.epa.gov/superfund/remedytech/tsp/download/gw_sampling_guide.pdf. Accessed May 2013.

Young, R.A., ed. 1993. *The Rietveld Method*. International Union of Crystallography Monographs on Crystallography, Vol. 5. Oxford: Oxford University Press.

Younger, P.L. 2007. *Groundwater in the Environment: An Introduction*. Oxford: Blackwell Publishing.

Zhang, C. 2007. *Fundamentals of Environmental Sampling and Analysis*. New York: John Wiley and Sons.

CHAPTER 4

Decision Making, Risk, and Uncertainty

Carol C. Russell, Kathleen S. Smith, and Virginia T. McLemore

4.1 INTRODUCTION

This handbook places special emphasis on answering of questions by gathering reliable scientific data and analyzing the data to provide information upon which to base decisions. Risks to human health and the environment are the underpinning for each of the identified international guidance documents for sampling and monitoring. Identifying risks to valuable resources and reducing uncertainty and doubt about alternatives allow a decision maker reasonable choices among alternatives.

This chapter introduces the principles of risk assessment and risk analysis, and their usefulness to the management of risk. Different types of uncertainty are described, including their importance to decisions that might be influenced by, or concerns with, management of the environment.

The Acid Drainage Technology Initiative–Metal Mining Sector (ADTI-MMS) recommends the step-by-step framework shown in Figure 4.1 for the development and implementation of a sampling and monitoring program.

A review of international guidance documents in English on sampling and monitoring programs for metals at mining and processing sites reveals the importance of identifying risks and their relevance for decision making. These guidance documents also provide recommended processes for quality assurance and quality control (QA/QC) to reduce uncertainty. Systematic steps in guidance documents are similar across jurisdictions and scopes of influence.

Recent emphasis on sustainability influences how the mining industry approaches environmental assessment, planning, and reporting (Fairbrother et al. 2007). For example, the International Council on Mining and Metals (ICMM) sustainable development framework (ICMM 2013) outlines approaches for biodiversity risk assessment and mitigation during the various phases of mining, and provides guidance for best management practices. Another example is the Global Reporting Initiative (GRI) reporting framework, which is intended to provide a generally accepted framework for reporting on an organization's economic, environmental, and social performance. The G3 is the third generation of the GRI reporting guidelines, and the GRI has a G3 *Mining and Metals Sector Supplement* (GRI 2010) that is designed for the mining and metals sector. This kind of sustainability reporting is a form of value reporting whereby an organization communicates their environmental, social, and economic performance, and can influence the need for the collection of specific types of data that have not traditionally been collected (ICMM 2008).

Generally, mining and processing operations throughout the world are expected to have individual and specific sampling and monitoring programs designed to address a variety of questions and issues arising during the mine life cycle. For example, monitoring of baseline water quality

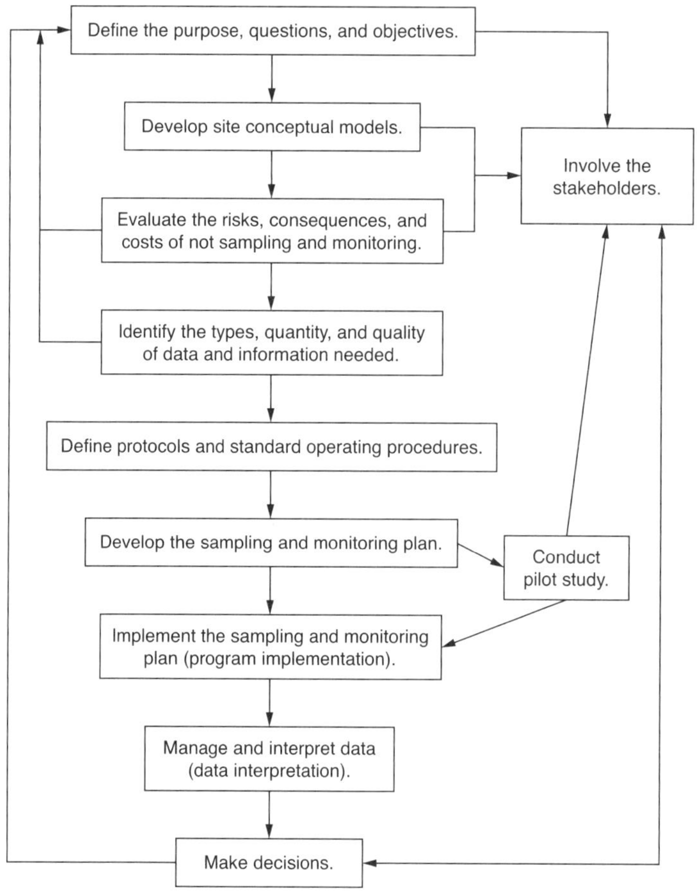

Source: Adapted from McLemore et al. 2004.

FIGURE 4.1 Steps for implementing a sampling and monitoring program

prior to initiating extraction may form the basis of future reclamation expectations and potential allocation of liabilities. Modeling of groundwater flow based on hydrogeological monitoring during the development phase could be used to predict impacts to groundwater resources at closure. Similarly, chemical analyses of exploration core and cuttings could be utilized to decide if special handling of waste rock will be necessary to ensure adequate reclamation.

4.2 RISK

The objective of sampling and monitoring is to provide adequate data to test or quantify a hypothesis to address potential environmental risks in a cost-effective manner and to minimize potential liabilities. "Mining is a high risk venture in view of the multitude of unpredictable factors in the finding, acquisition, extraction, treatment and sale of the product. Risks may be geological, technological, economic, political and/or social" (Lacy 2010). Throughout the mine life cycle, from the decision to invest in a new mine, to the safety and health issues of working or living near a mine, to the environmental issues during production and ultimate closure and the potential future land uses, there are risks. In this chapter, an attempt is made to bridge the gap between different concepts of risk as used by professionals operating in the mining, environmental, and economic realms.

Severity of Effect	Probability of Effect: A Almost Certain	B Likely	C Possible	D Unlikely	E Rare
1 Catastrophic	Extreme	Extreme	Extreme	Extreme	High
2 Major	Extreme	Extreme	Extreme	High	High
3 Moderate	Extreme	High	High	Moderate	Moderate
4 Minor	High	High	Moderate	Low	Low
5 Insignificant	High	Moderate	Low	Low	Low

Source: Adapted from Australian Government 2008.

FIGURE 4.2 Classification of risks associated with mining operations in South Australia (the darker the shading, the higher the risk)

GEORGIUS AGRICOLA

Georgius Agricola says this about the risk of mining:

> *The critics say further that mining is a perilous occupation to pursue because the miners are sometimes killed by the pestilential air which they breathe; sometimes their lungs rot away; sometimes the men perish by being crushed in masses of rock; sometimes falling from ladders into the shafts, they break their arms, legs, or necks; and it is added, there is no compensation which should be thought great enough to equalize the extreme dangers to safety and life.*
>
> —*De Re Metallica*

The term *risk* means different things to different people. Risk is usually seen as an uncertain event that could adversely affect the achievement of objectives. Risk is a function of the consequences of an event, such as loss of life or economic damages, combined with the likelihood of its occurrence (Figure 4.2). To take Agricola's example (see sidebar) of falling off ladders, an objective risk assessment uses numbers, after the fact, to calculate the risk. In contrast, by proactively estimating potential risks in the future, alternatives can be evaluated based on projected environmental impacts and mitigation costs to design a mine plan and to set an adequate environmental bond.

Metals and metal compounds have unique characteristics that should be considered when assessing their risks. For example, metals are neither created nor destroyed by biological or chemical processes, but they can be transformed from one chemical form to another. Metals occur naturally, and some metals are essential for life in low doses but are toxic at high levels. The *Framework for Metals Risk Assessment* (USEPA 2007, 2011) is an excellent place to start to assess metals in the environment. In environmental science, the study of adverse impacts originating from a source is commonly considered to be the study of toxicology or the science of poisons. These adverse effects may occur in many forms, ranging from immediate death (acute toxicity) to subtle changes not realized until months or years later (chronic toxicity).

4.2.1 Risk Assessment

Risk assessment is typically defined as a measure of the probability of an event happening and severity of adverse effects. Risk assessment answers the following questions (Kaplan and Garrick 1981; Suter 1993):

- What can go wrong?
- What is the likelihood that it would go wrong?
- What are the consequences?

Answers to these questions help risk analysts to identify, measure, quantify, and evaluate risks and their consequences and impacts. "A successful risk assessment leading to a reduction or elimination of risk is only as good as the knowledge, skill, and effort of those undertaking the assessment" (Caldwell 2010). One also must take into account public perceptions of risk and attempt to manage potential public misunderstandings. Risks considered most serious by the general public, in many cases, are different from those considered most serious by the technical professionals charged with reducing environmental risk. Consideration of public perceptions about selected chemicals associated with the mining and processing industry, such as cyanide leaching during processing, should be considered as a monitoring program is designed.

A human health risk assessment (HRA) consists of the evaluation of scientific information on the hazardous properties of environmental agents (hazard characterization), the dose-response relationship (dose-response assessment), and the extent of human exposure to those agents (exposure assessment). In final form, a risk assessment is a statement regarding the probability that populations or individuals so exposed will be harmed and to what degree (risk characterization). Especially vulnerable populations, such as children, are often taken into account. The two major databases used for comparison are the Integrated Risk Information System (IRIS) in the United States and Registration, Evaluation, and Authorization of Chemicals (REACH) by the European Chemicals Agency. The IRIS database contains information for more than 540 chemical substances on human health effects that may result from exposure to various substances in the environment, whereas the REACH system is more of a voluntary registration of 8,469 unique, manufactured substances. Both rely on the concept of a dose-response curve with a *safe region* of no effects transitioning to an increasing *not safe* zone. Noncancer risks are assumed to have a threshold, but cancer risks have no threshold.

Metal-specific tools such as the *Framework for Metals Risk Assessment* (USEPA 2007, 2011) in the United States and the *Health Risk Assessment Guidance for Metals* (HERAG; ICMM 2006) and *Metals Environmental Risk Assessment Guidance* (MERAG; ICMM 2007) each provide internationally relevant scientific references for the application of risk assessment to metals with regard to human health and the environment. Risk assessments conducted on specific metals in the European Union and North America and these national guidance documents acknowledge that metals have specific properties that require consideration in risk assessment. The main HERAG and MERAG documents provide an overview of the particular issues associated with the evaluation of metals. A series of associated fact sheets address selected issues in more detail (ECHA 2008). ICMM has provided a comparison of the generic risk management process and the generic environmental assessment framework (ICMM 2007). Table 4.1 provides a comparison of the risk assessment process from the Management Advisory Board Australia and the generic environmental assessment framework for U.S. Environmental Protection Agency's (USEPA's) ecological risk assessments.

TABLE 4.1 Comparison of the generic risk assessment process

Step	Management Advisory Board Australia Process	Generic Framework	USEPA Framework for Ecological Risk Assessment
1	Establish the context	Define policy, purpose, objectives, success criteria	Problem formulation
2	Identify the risks	Concerns, consequences	Problem formulation
3	Analyze the risks	Calculations (identify concerns and possible outcomes); certainty and uncertainty	Analysis phase
4	Assess and prioritize risks	Compare with criteria; prioritize	Risk characterization
5	Treat the risks	Control, communication; develop and implement contingency/ management plans	Communication
6	Review and monitor	Continue reviewing and monitoring; assess effectiveness of treatment	Review and monitor; assess effectiveness of treatment

Source: Compiled by ICMM (2008), as adapted from Beer and Ziolkowski (1995).

TABLE 4.2 Comparison of terminology used in sampling and monitoring with that used in risk assessment and risk management

Sampling and Monitoring	Risk Assessment	Risk Management
Source—Geoavailability	Hazard identification	Alternative development (options for releases)
	Exposure assessment	Evaluation of alternatives
Pathway	Dose-response	Mitigation
		Recommendations
Receptor	Toxicity assessment	Action (implementation)

4.2.2 Risk Management

Risk management is commonly distinguished from risk assessment in that it focuses on what can be done regarding what has or what may happen (see Table 4.2 and Chapter 5). Risk management builds on the risk assessment evaluating the following:

- What can be done and what options are available?
- What are the associated trade-offs in terms of all costs, benefits, and risks?
- What are the impacts of current management decisions on future options?

Risk management is the assessment and prioritization of risks (defined in ISO 31000 as the effect of uncertainty on objectives, whether positive or negative), followed by coordinated and economical application of resources to minimize, monitor, and control the probability and/or impact of unfortunate events or to maximize the realization of opportunities. Strategies to manage risk include transferring the risk to another party, avoiding the risk, reducing the negative effect of the risk, and accepting some or all of the consequences of a particular risk (Failing et al. 2007). Figure 4.3 illustrates the process for risk assessment and management for metals (which is discussed in more detail in USEPA 2007, 2011).

4.2.3 Environmental Monitoring—What Is at Risk?

At risk are human health and the environment as reflected by the quality of air and water, ecosystem health, and other ecosystem services and values such as endangered species, and visual, material, and historical assets. Risk assessment and risk management should be tied to the mine life cycle and take a comprehensive, integrated look at mining and mineral processing procedures,

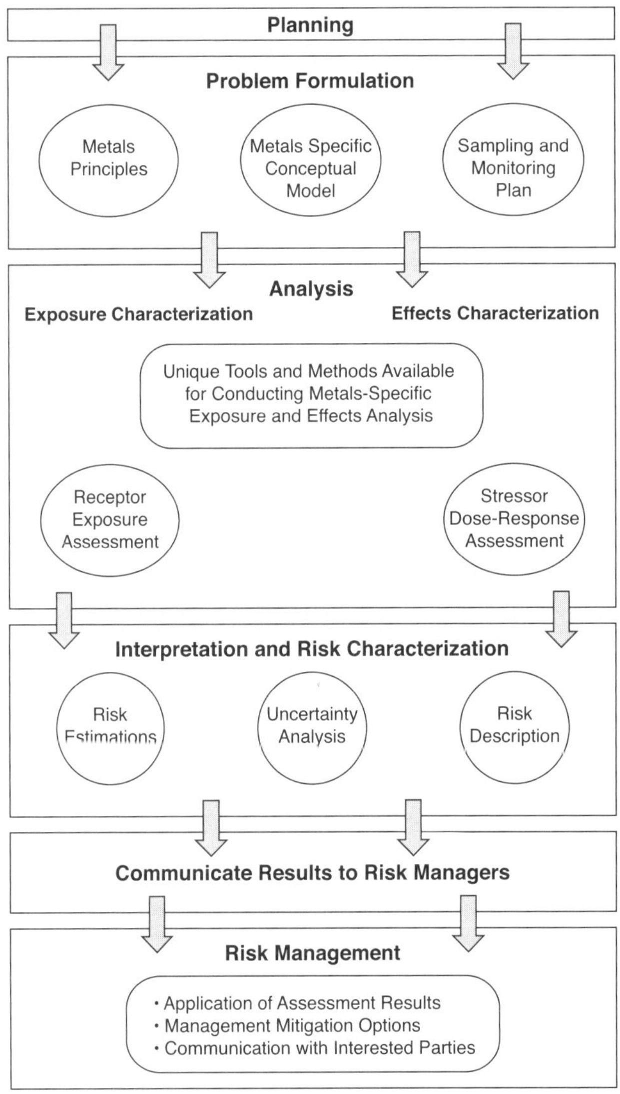

Source: USEPA 2011.

FIGURE 4.3 Risk assessment/risk management process for metals

materials produced, by-products, and final products. By integrating mining and mineral processing with an environmental risk assessment, proactive risk management decisions can be made to reduce environmental liabilities and to protect the environment. Some mine sites are large complexes that can have multiple sources of potential concern besides the mining and processed materials, and can include chemicals used in mining and processing, septic wastes, explosives, electrical transformers and power systems, fuel, lubricants, cleaners, and laboratory wastes. The HRA process includes estimating the quantity of contaminants from a site reaching human populations and determining how toxic these contaminants are to humans. In a similar fashion, the environmental risk assessment proceeds from the source via a pathway and results in a toxicity to the receptors. Risk management is the targeting of protective efforts or mitigation to the problems

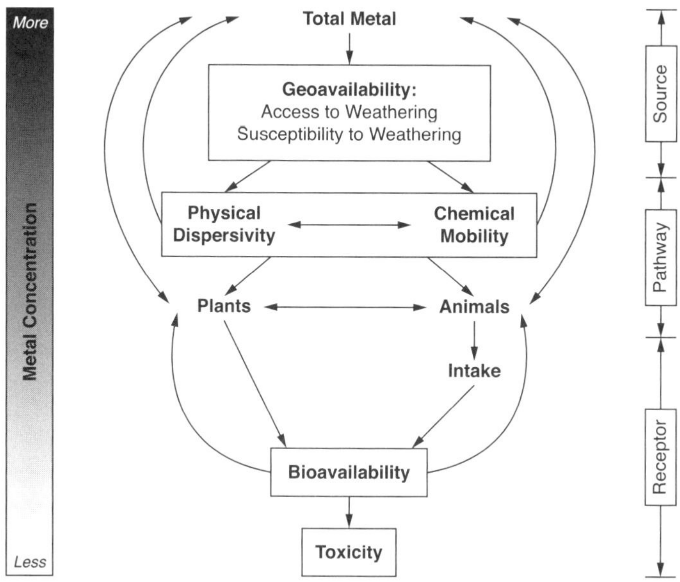

Source: Adapted from Smith and Huyck 1999; Smith 2007.

FIGURE 4.4 Source-pathway-receptor risk diagram

that seem to be the most serious or that appear to provide the most protection or mitigation for the effort expended.

A metal is only toxic if it is in a form and concentration that is bioavailable at the time of exposure (Figure 4.4; Smith and Huyck 1999; Smith 2007). Risk assessors use a variety of methods to determine metal bioavailability in aquatic systems. These methods include adjusting for water hardness, water-effect ratio (see Appendix 3), and biotic ligand models (BLMs; USEPA 2007, 2011; see Appendix 3). BLMs combine the influences of metal speciation and cationic competition on aquatic metal toxicity. Bioconcentration factors (BCFs) and bioaccumulation factors (BAFs) are commonly used to quantify metal accumulation in tissue relative to the water concentration. In the case of metals at mining and processing sites, care must be used in applying BCFs and BAFs due to the complexities involved in determining relationships between metal speciation and bioavailability, and in direct versus indirect metal toxicity (Chapman et al. 1996).

4.2.4 Risks and Consequences

Consequences of poor site characterization can be serious. In particular, water availability and water quality issues have the potential to significantly impact a company's financial performance. In general, water-related risks can be broadly classified into the categories shown in Table 4.3.

Mining companies have long been conscious of water-related risks, as evidenced by their ongoing efforts to address them and related corporate reporting (Barton 2010). Indeed, recent analysis has shown that the mining sector is a leader in terms of water quality reporting for the GRI (2010). However, corporate disclosure often does not provide a comprehensive picture of water-related risk. Current reporting frameworks do not guide companies to disclose the full scope of potential water-related risks. *Mine the Gap: Connecting Water Risks and Disclosure in the Mining Sector* by Miranda and Sauer (2010) of the World Resources Institute outlines potential water-related risks facing the mining industry and highlights important gaps in water-related disclosure. The purpose is to provide information, questions, and tools to help the financial

TABLE 4.3 Water availability and quality risks for mining

Risks	Company's Financial Perspective
Physical	
Losses due to production	Lost revenues
Losses due to lower mineral recovery	Lower margins, lost revenues
Losses due to degraded mineral quality	Lower margins, lost revenues
Losses due to conflicts with communities	Disruption, lost revenues
Regulatory and Legal	
Losses due to operating costs	Lower margins, lost revenues
Capital expenditures	Reduced cash flow, lost revenues
Losses due to production	Lost revenues
Mine shutdown	Lost revenues, value destruction
Closure liabilities	Valuation
Reputational	
Operation disruptions	Lost revenues
Lost access to future reserves	Value destruction, lost revenues
Access to capital	Cost of capital

Source: Adapted from Miranda and Sauer 2010.

community better evaluate water-related risks facing mining companies (Miranda and Sauer 2010), specifically:

- Water-related risks span the minerals production cycle and occur in diverse operating environments.
- Water quality data are not sufficiently reported. Water quality problems are among the most serious environmental impacts associated with mining. Data on water effluents and waste management practices are either not reported or not detailed enough to understand risk. Toxic waste and mine effluents can be mobilized by water, resulting in regulatory, legal, and reputational risks for companies (see Kuipers et al. 2006 for examples).
- Work stoppages at mine or processing facilities can occur if water resources become unavailable. Mining and processing—particularly for precious metals, diamonds, copper, and nickel—requires significant volumes of water (Miranda and Sauer 2010).

4.2.5 Costs Within a Mining Risk Analysis Framework

Evaluation of the costs and potential consequences of not sampling and monitoring can be considered a type of cost/benefit evaluation or mining operation risk analysis (Figure 4.5). Throughout the mine life cycle, balancing the costs and benefits of monitoring is advised. For example, monitoring of background and baseline conditions during the exploration phase can provide the first important assessment of whether the mine site can be remediated during operation and after closure. Monitoring during design can provide better operating procedures, which can minimize environmental and bonding costs. Monitoring during mining and processing operations can be used in process control to identify problems before they become major issues and while there is still a positive cash-flow to mitigate such issues. Finally, postclosure monitoring can minimize long-term costs and expedite bond release.

Most companies strive, for obvious economic reasons, to obtain the maximum benefit at minimum cost, meaning that the number of drillholes must be just sufficient to ensure definition

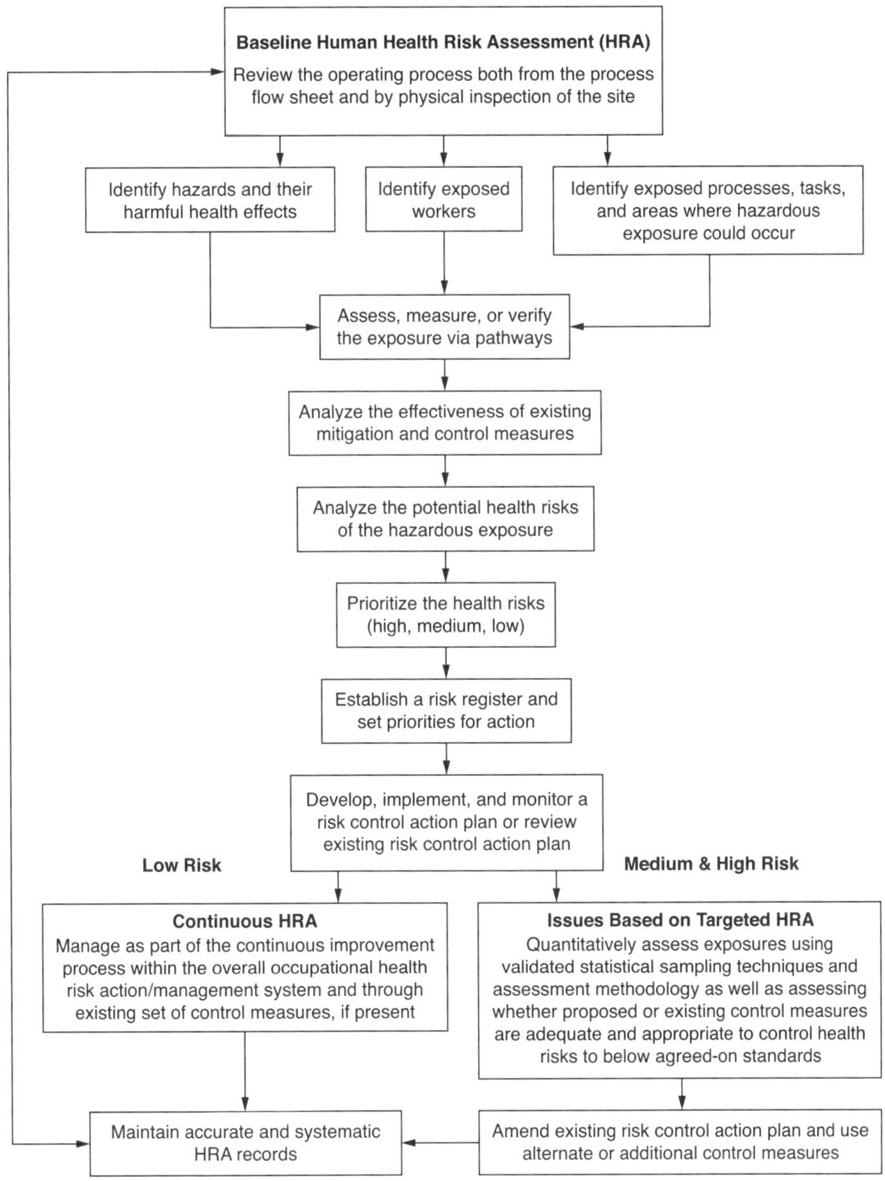

Source: Adapted from ICMM 2008.

FIGURE 4.5 Health risk assessment cycle for mining operations

of the ore zone without costing more than necessary. The inevitable question that arises is *What is the optimum drillhole spacing?* The answer is *it depends!* The level of risk that management is willing to accept determines how rigorous data collection will be at the start of a project as well as during subsequent mine development. The cost of gathering information has to be weighed against the planning and operational risk to a mining company based on the reliability of the data compared to the importance or value of the ecological, regulatory, or economic component being studied. The *Handbook of Systems Engineering and Management* (Rouse 2009) describes the issue very well. (See also Engineering Systems and Risk sidebar.)

ENGINEERING SYSTEMS AND RISK

"Engineering systems are almost always designed, constructed, and operated under unavoidable conditions of risk and uncertainty and are often expected to achieve multiple and conflicting objectives. The identification, quantification, evaluation, and trading off of risks, benefits and costs should constitute an integral and explicit component of the overall managerial decision-making process and should not be a separate, cosmetic afterthought." (Rouse 2009).

In mineral evaluation, risks to stakeholders, specifically stockholders, is a topic of great interest and concern. The mining and mineral processing community is very familiar with the concepts of data reliability. The level of confidence in key information is the distinction between proven and probable reserves. The figures are always estimates rather than precise calculations, with margins of error that should be recognized in the way the figures are expressed. In 2002, the CRIRSCO (Combined Reserves International Reporting Standards Committee, now known as the Committee for Mineral Reserves International Reporting Standards) was formed with the mission to continue coordination between member countries regarding international resources and reserves definitions. Considerable progress has been made toward widespread adoption of standards throughout the world. However, mining companies subject to U.S. Securities and Exchange Commission (SEC) regulations must comply with Industry Guide 7 when reporting ore reserves (SEC 2013). In 2012, the Society for Mining, Metallurgy and Exploration (SME) petitioned the SEC to update its Industry Guide 7 using as a model the 2007 SME Guide based on the CRIRSCO recommendations (SME 2007). As shown in Figure 4.6, the choice of the appropriate category of mineral resource depends on the quantity, distribution and quality of data available, the level of confidence that attaches those data, and the appropriateness of the estimation methodology applied. Those data include mining, metallurgical, economic, marketing, legal, environmental, social, and governmental factors, many of them with the potential to block a new mine or make serious problems for an established one.

As noted at the bottom of Figure 4.7, environmental factors are also taken into consideration. Specifically, the Canadian requirements address QA/QC and include the following environmental requirements (CIM 2010): baseline studies, tailings management, waste rock management, acid rock drainage (ARD) issues, closure and reclamation plans, and permitting schedule.

The 2007 SME Guide with international resources and reserves definitions or their precursors (i.e., Goudarzi 1984) were accepted as part of national reporting codes and guidelines by the regulatory agencies of Australia (1989), South Africa (2000), Canada (2001), and the United Kingdom (2001). The United Nations formally adopted these definitions in 1999. In 2007, the government of Chile approved a reporting code that includes these definitions. In 2008, the Pan-European Code was issued by the Pan-European Reserves and Resources Reporting Committee, and in 2011, the Russian Code was issued by the National Association for Subsoil Examination.

In a similar manner, planning and operational risks can be placed in perspective based on the value placed on the potentially impacted part of the environment and the reliability of the data. Environmental risk analyses for closure based on unreliable data can be underprotective or overprotective of ecosystems, with the result that there are added cost penalties on society and on industry to achieve these levels of protection. Some risks are potentially so serious, and the time for recovery so long, that risk reduction actions should be viewed as a kind of insurance premium

ENDANGERED SPECIES AT HENDERSON, COLORADO

Carol C. Russell, U.S. Environmental Protection Agency

An example of the complications encountered when an endangered species appears to prefer an acid rock drainage environment, such as a mine tailings pond, is illustrated by the Henderson mine in central Colorado. The boreal toad (*Bufo boreas boreas*), listed as a State Endangered Species and a candidate for federal protection under the Endangered Species Act, only breeds in still or slow-moving acidic water such as the water found in a tailings impoundment. The mine took on the toad as their mascot and actively participated in the efforts to establish new (nonmine) habitat (Russell 2006).

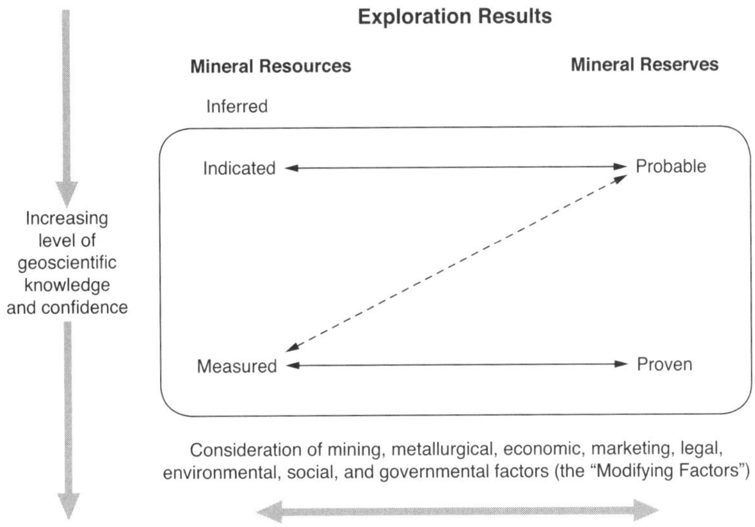

Source: SME 2007.

FIGURE 4.6 General relationships between exploration results, mineral resources, and mineral reserves

and initiated in the face of incomplete and uncertain data. The risks entailed in postponing action can be greater than the risks entailed in taking inefficient or unnecessary action. In short, the only way to decrease one's exposure to risk is to increase the reliability of the data or to decrease the risk using risk management practices. Other complications can enter into the decision of if, when, and how to remediate an area. For example, Leadville, Colorado, depends on the tourist industry for much of its economy, and the complete removal of waste rock piles would detract from the visual effect of the mining town. Therefore, special coverings of the waste rock piles were designed to protect the visual effect as well as reduce the ARD. In some areas, endangered and other species depend on mining influenced water and must be protected, such as at the Henderson mine in Colorado (see Endangered Species sidebar).

Several strategies or processes are involved in risk analysis, as shown previously in Figure 4.3. The HRA cycle for mining operations is shown in Figure 4.5 (ICMM 2008), although a simpler approach combining risk assessment with risk management is provided by Dowd (2002):

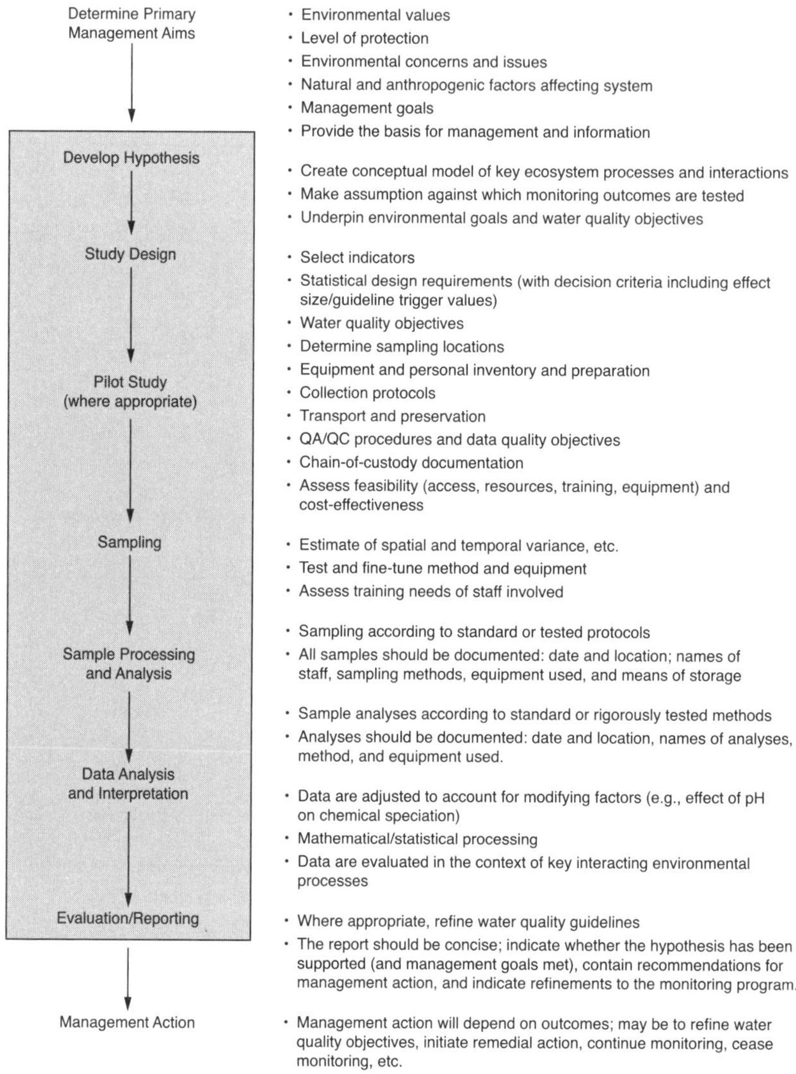

Source: Adapted from the Monitoring and Reporting Guidelines and the framework for designing a wetland monitoring program adopted by the Ramsar Wetland Convention (Ramsar Convention 1996; Finlayson 1996; ANZECC and ARMCANZ 2000).

FIGURE 4.7 Procedural framework for the monitoring and assessment of water quality

1. Identify the risk and the sources.
2. Assess and quantify the risk, including assessment of the exposure, pathways, and transport modes.
3. Determine the minimum acceptable level of risk and the consequences of the risk.
4. Manage the residual risk by developing proper sampling and monitoring plans.
5. Reduce the risk to a minimum for adequate quality assurance and quality control.

4.3 INTERNATIONAL GUIDANCE DOCUMENT COMMONALITIES

Many countries and organizations have endorsed a data quality objectives structure for planning a sampling and monitoring program as described in Table 4.4. The following sections provide an overview of frameworks in various jurisdictions and by different organizations. The basic components of the international guidance documents are very similar based on the information readily available in electronic form. Much more information is available from listed references.

4.3.1 Australian and New Zealand Sampling and Monitoring Programs

The Australian and New Zealand monitoring framework focuses on risk management by protecting water quality, rather than focusing on development of sampling and monitoring guidelines. The Australian guidelines also provide risk-based decision frameworks to help the user refine guideline trigger values for application at local and/or regional scales. The guidelines initially used a very deterministic view of ecosystems, assuming that factors controlling ecosystem function could be identified and managed to prevent problems. However, now they are more inclusive of the complexity of mining and processing activities as well as potential impacts to receptors. Monitoring includes both observation (e.g., recording information about the environment) and investigation (e.g., manipulative studies such as toxicity tests where environmental conditions are controlled). Although there is great reliance on the use of biological indicators, all monitoring relies on sound practice in environmental science, as shown in Figure 4.7, including

- Explicit written definition of the sampling site, project objectives, a hypothesis, and the sampling protocol that will support the work;
- The definition of sampling sites, sampling frequency, and spatial and temporal variability that will permit appropriate statistical methods to be used;
- Rigorous attention to field and laboratory quality control and assurance;
- Incorporation of a pilot study to test the sampling protocol and determine spatial and temporal variability; and
- Mitigation measures.

Specifically, the Australian and New Zealand guidelines reference procedures for determining an unacceptable level of change (ANZECC and ARMCANZ 2000).

Environment Australia also developed an internal QA/QC document for the Australian River Assessment System in New South Wales. This was undertaken to ensure that only QA data were used in all aspects of model development, model testing, and river site assessment addressing issues of lineage, positional accuracy, attribute accuracy, logical consistency, and completeness. Past data collection and assessment demonstrated that a screening process on data was needed to ensure that it was adequate to make decisions on river conditions (Waddell 2000).

4.3.2 European Union's Sampling and Monitoring Program

The European Union water framework directive guidance provides exceptional insight and diagrams of basic questions that need exploration before a monitoring program is implemented (European Communities 2003). The directives for water and waste are media-focused based on changes to the quality of the water, soil, or air (Figure 4.8). For water, the directive provided guidance not only for surface water but also for groundwater and wetlands. However, specific monitoring guidance for mining-related sampling of water was limited.

The waste framework directive, in contrast, has provided a great deal of research on this subject of monitoring mining; specifically, several guidance documents regarding characterization of waste from extractive industries, including

- Static testing for determination of acid potential and neutralization potential of sulfidic waste,

TABLE 4.4 International sampling and monitoring elements guidance document comparison

	ADTI-MMS*	Australia and New Zealand†	European Union‡	Canada§	USEPA**	GARD Guide††
1.	Define the questions and objectives	Defines values, goals, water quality objectives, and level of protection	Issues for consideration: objectives, variability	Define the project background and goals	State the problem; identify the goals	Objectives of monitoring
2.	Develop site conceptual models	Determine appropriate guidelines—local environmental conditions	Development of a conceptual understanding	Develop a conceptual site model	Define the boundaries of the study	Conceptual and dynamic system model development
3.	Evaluate the risks, consequences, and costs	Define specific water quality to be achieved	Acceptable level of risk, precision, and confidence	Establish the investigation objectives		Risk-based approach to monitoring
4.	Identify the types, quantity and quality of data, and information needed				Identify the information inputs and analytic approach; specify performance criteria	
5.	Define protocols	Establish monitoring and assessment program after defining acceptable performance or decision criteria	Quality assurance/ quality control, standard methods			
6.	Develop the sampling and monitoring plan	Monitoring plans are left to the local agencies		Prepare a sampling and analysis plan	Develop the plan for collecting the data	Monitoring program development
7.	Conduct the sampling and monitoring plan	Monitoring is undertaken on a case-by-case basis	Implementation of quality assurance programs	Conduct the field investigation program		
8.	Analyze and manage data	Initiate management response; "Trigger values"	Expression of results	Validate and interpret the data		Data management and interpretation
9.	Decision making	Stakeholder process	Transparency of process			Auditing

* ADTI-MMS (Acid Drainage Technology Initiative–Metal Mining Sector; McLemore et al. 2004, 2007, 2009).
† Australian and New Zealand Environment and Conservation Council (ANZECC and ARMCANZ 2000).
‡ European Communities (2003); European Commission (2012).
§ Environment Canada (2004, 2005, 2012, 2013); MEND (Mine Environment Neutral Drainage; www.mend-nedem.org).
** USEPA (2006, forthcoming).
†† *Global Acid Rock Drainage Guide* (INAP 2009).

Biological

- Plants (abundance, diversity, composition, taxa)
- Macro-Invertebrates (abundance, diversity, composition, taxa)
- Phytoplankton (abundance, diversity, composition, taxa)
- Fish (abundance, diversity, composition, taxa)

Nonsynthetic Pollutants

- Nutrients (priority list substances)
- Metals (priority list substances)

Synthetic Pollutants

- Organics (priority list substances)
- Inorganics (priority list substances)

Hydromorphological

- Waterbody Class (surface water, groundwater, wetlands)
- Flow/Depth/Velocity
- Morphological Condition (continuity, channel pattern, riparian zone, substrate)
- Seasonality

Physicochemical

- pH
- Salinity/Total Dissolved Solids
- Temperature
- Oxygen
- Clarity/Turbidity/Suspended Solids
- Nutrient conditions

Source: Adapted from European Communities 2003.

FIGURE 4.8 European Commission water framework—selection of quality elements for transitional waters

- Kinetic testing for assessing acid generation potential of sulfidic waste from extractive industries,
- An overall guidance document for characterization of waste from extractive industries, and
- Sampling and analysis of weak acid dissociable cyanide discharged into tailings ponds.

4.3.3 Canadian Sampling and Monitoring Program

The MEND (Mine Environment Neutral Drainage) program in Canada is known for its cooperative research program between Canadian mining companies and provincial/territorial and federal departments. Since 1989, MEND has worked to develop technologies to prevent and control acidic drainage from mining operations. The program is directed by a multistakeholder committee, with members from the mining industry, federal and provincial governments, and nongovernment organizations. Specific publications on sampling and monitoring include the following categories: (1) prediction, (2) prevention and control, (3) treatment, (4) monitoring, and (5) technology transfer. Publications of particular note are

- *Field Sampling Manual for Reactive Sulphide Tailings* (CANECT Environmental Control 1989),
- *Review of Waste Rock Sampling Techniques* (SENES Consultants Limited et al. 1994),
- *Handbook for Waste Rock Sampling Techniques* (SENES Consultants Limited 1994),
- Appendix A: Technical Summary Note in *Guideline Document for Monitoring Acid Mine Drainage* (Terrestrial and Aquatic Environmental Managers and SENES Consultants Limited 1997), and
- *Application of Remote Sensing and Geophysics to the Detection and Monitoring of Acid Mine Drainage* (Paterson, Grant and Watson Limited and Geomatics International, Inc., 1994).

In addition, in 2012, Environment Canada published *Metal Mining Technical Guidance for Environmental Effects Monitoring*. Chapter 2, titled "Study Design, Site Characterization and General Quality Assurance/Quality Control," mainly focuses on aquatic effects to fish of mining in Canada, but it does provide for data quality measures derived from intended data uses (e.g., hypotheses to be tested, summary statistics involved, and total uncertainty that can be tolerated; Environment Canada 2012). Section 2.2.1, Site Characterization, in that document focuses on specific media (soil, water, air) and specific contaminants within that media. The Canadian section on conceptual models is particularly useful in that details of different sites are provided as examples. Also, the document provides detailed checklists for site assessment leading to a risk assessment. The Canadian Soil Quality Guidelines for the Protection of Environment and Human Health summary tables (CCME 2007) are being revised.

British Columbia has also been very active on guidance for the environmental effects of mining with the recent *Water and Air Baseline Monitoring Guidance Document for Mine Proponents and Operators* (Ministry of Environment 2012). This guidance document outlines and defines the baseline study requirements and information considerations necessary to propose a mineral development project in the province of British Columbia. By providing these requirements to the proponent early in the project evaluation stage (exploration), the Ministry of Environment aims to promote effective study design and duration and to ensure that information collection and data usage/interpretation will assist in the initial project evaluation and be useful throughout the development, operation, and closure of a mine.

4.3.4 USEPA Sampling and Monitoring Program

The USEPA sampling and monitoring program has served as the framework for numerous other international environmental QA/QC guidance documents. The USEPA Quality System (the sampling and monitoring program) is based on developing performance and acceptance criteria, or data quality objectives (DQOs), and qualitative and quantitative statements that define the purpose for the data collection effort, clarify the kind of data needed, and specify the limits on decision errors needed for the study.

A sampling and monitoring program can be composed of several steps to document the process of QA/QC (see Section 3.13). A goal of USEPA's Quality System is to ensure that environmental programs and decisions are supported by data of the type and quality needed and expected for their intended use. The DQO process is a series of steps that serve as a guide for the resource-effective acquisition of environmental data and apply to both decision making and estimation. It is used to develop performance and acceptance criteria that clarify the objectives of the investigation, define the appropriate type of data, and specify tolerable levels of potential decision errors that will be used as the basis for establishing the quality and quantity of data needed to support decisions (USEPA 2006). The Quality System (see USEPA 2006) includes a series of guidance documents that relate to one another, as shown in Figure 4.9.

The quality assurance project plan (QAPP) describes the policy, organizational, and functional activities necessary to collect data that will stand up to legal and scientific scrutiny (USEPA 2002a). The quality management plan describes processes and procedures at the organizational level, management and staff functional responsibilities, and line of authority in the USEPA guidance QA/R-2 (USEPA 2001c). The QAPP (USEPA 2002a) is the document that details the who, what, when, where, why, and how of the project objectives (DQOs) found in QA/R-5 (USEPA 2001b) and QA/G-5 (USEPA 2002b). The field sampling plan provides the field sampling details for the sampling event. The sampling and analyses plan provides the field and analytical details for the sampling event(s) and specifies the process for obtaining environmental data of sufficient quality to satisfy the project objectives. Sampling and monitoring plans are discussed further in Chapter 6, and examples are provided in Appendix 4. Standard operating procedures describe

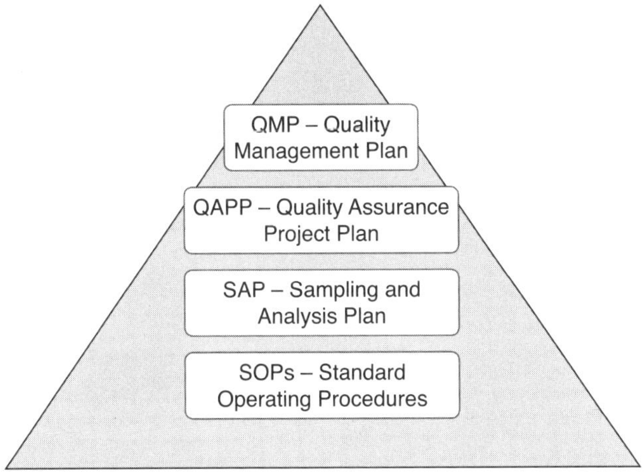

Courtesy of Mary P. Goldade.

FIGURE 4.9 USEPA's Quality System

specific sampling techniques, field analytical methodologies, or analytical methods, many of which are described in Appendix 3.

Three guidance documents from the United States focus on metals and mining sampling and monitoring:

1. *Framework for Metals Risk Assessment* (USEPA 2007, 2011)
2. *EPA and Hardrock Mining: A Source Book for Industry in the Northwest and Alaska* (USEPA 2003)
3. *Abandoned Mine Site Characterization and Cleanup Handbook* (USEPA 2001a)

In addition, Karl Ford, formerly with the U.S. Department of the Interior Bureau of Land Management, wrote the book *Risk Management Criteria for Metals at BLM Mining Sites* (Ford 2004). Furthermore, the U.S. Geological Survey has numerous publications on the subject both through the U.S. government press and in the peer-reviewed literature (see http://crustal.usgs.gov/projects/minewaste/).

4.3.5 *Global Acid Rock Drainage Guide*

The *Global Acid Rock Drainage Guide* (GARD Guide) is a Web-based, actively updated resource intended as a state-of-practice summary of the best practices and technology dealing with the prediction, prevention, and management of drainage produced from sulfide mineral oxidation, often termed *acid rock drainage*, *saline drainage*, *acid mine drainage* or *acid and metalliferous drainage*, *mining influenced water*, and *neutral mine drainage*. It was developed on behalf of the International Network for Acid Prevention, an international consortium of mining companies, by consultants, academicians, mine operators, and regulators to address issues related to sulfide mineral oxidation. In Chapter 8 of the GARD Guide, *monitoring* is defined as the process of routinely, systematically, and purposefully gathering information for use in management decision making. Mine site monitoring characterizes environmental changes from mining activities to assess conditions on the site and possible impacts to receptors. Monitoring includes both observation (e.g., recording information about the environment) and investigation (e.g., manipulative studies such as toxicity tests where environmental conditions are controlled). An example of the use of monitoring in

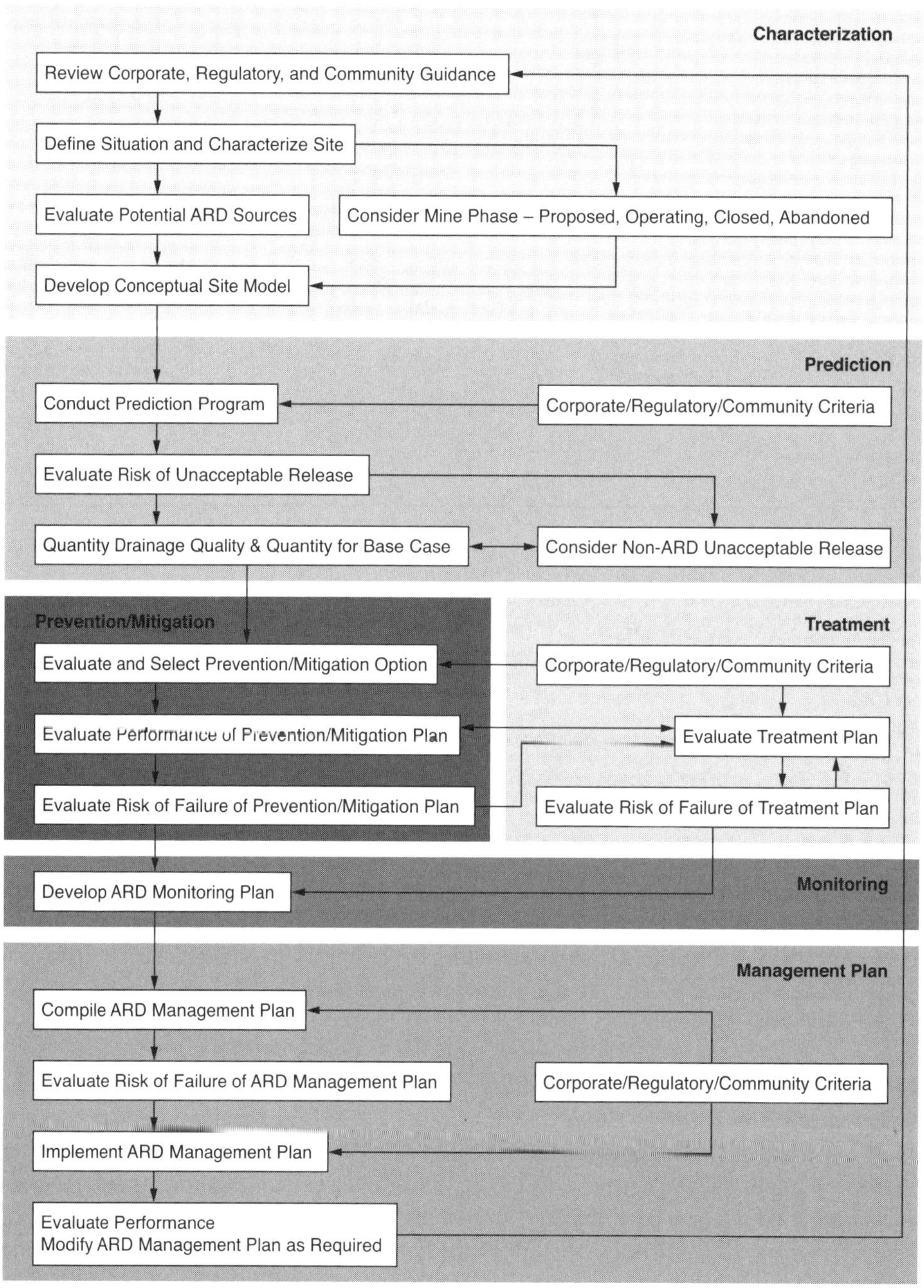

Source: INAP 2009.

FIGURE 4.10 GARD Guide overall management plan

management decision making was provided to assess the effectiveness of mitigation measures to minimize the effects of environmentally detrimental processes such as mining influenced waters and implementation of adjustments to mitigation measures. Figure 4.10 is a flow diagram of the incorporation of monitoring into a mining company's decision-making process.

4.4 SUMMARY

In summary, this chapter is not proscriptive but is designed as a thought-provoking method to guide users in articulation, investigation, analysis, and documentation to provide sufficient, quality information to decision makers. In designing a monitoring program, the objectives of the sampling effort must be determined first. What questions need to be answered in order to make decisions? How detailed must the data be to answer the questions? The ability to assess environmental risks, compare them, and select strategies to reduce them all depend on the availability and sophistication of the relevant data and analytical tools. Based on the assessment of risks, the sampling program can be designed considering the necessary spatial and temporal resolution, and the number of samples and replicates needed to satisfactorily address the study objectives.

Documentation of uncertainties or limitations of the data or assumptions made are as important as the range of possible risks. The quality of the data necessary to answer the questions is based on the importance or value of the ecological, regulatory, or economic issues at risk. In fact, some long-term and widespread environmental problems should be considered relatively high risk even if the data on which the risk assessment is based are somewhat incomplete and uncertain. The risks entailed in postponing action can be greater than the risks entailed in taking inefficient or unnecessary action.

International guidance documents follow the same general format for implementation of a decision-based sampling and monitoring program. ADTI-MMS tailored these guidance documents for mining and mineral processing in an effort to standardize expectations in the same manner as the international minerals evaluation processes are coming together.

4.5 REFERENCES

ANZECC and ARMCANZ (Australian and New Zealand Environment and Conservation Council and Agriculture and Resource Management Council of Australia and New Zealand). 2000. *Australian and New Zealand Guidelines for Fresh and Marine Water Quality*. National Water Quality Management Strategy No. 7, Paper No. 4. www.environment.gov.au/water/publications/quality/pubs/nwqms-guidelines-4-vol1.pdf. Accessed 9/11/13.

Australian Government. 2008. *Risk Assessment and Management*. Commonwealth of Australia www.ret.gov.au/resources/Documents/LPSDP/LPSDP-RiskHandbook.pdf. Accessed 9/26/13.

Barton, B. 2010. *Murky Waters? Corporate Reporting on Water Risk*. Boston: Ceres. www.ceres.org/resources/reports/corporate-reporting-on-water-risk-2010. Accessed 9/16/13.

Beer, T., and Ziolkowski, F. 1995. *Environmental Risk Assessment: An Australian Perspective*. Supervising Scientist Report 102. www.environment.gov.au/ssd/publications/ssr/pubs/ssr102-printquality.pdf. Accessed 9/11/13.

Caldwell, J. 2010. Learning to live with risk. Mining.com. www.mining.com/2010/07/01/learning-to-live-with-risk/. Accessed April 2011.

CANECT Environmental Control. 1989. *Field Sampling Manual for Reactive Sulphide Tailings*. MEND Project 4.1.1. www.mend-nedem.org/reports/files/4.1.1.pdf. Accessed 9/11/13.

CCME (Canadian Council of Ministers of the Environment). 2007. Canadian soil quality guidelines for the protection of environmental and human health: Summary tables. Updated September 2007. In Canadian Environmental Quality Guidelines (1999). Winnipeg: CCME. www.popstoolkit.com/Tools/SitePrioritization/Files/Guidelines/SoilQ%20eco%20and%20human%20h.pdf. Accessed 12/20/13.

Chapman, P.M., Allen, H.E., Godtfredsen, K., and Z'Graggen, M.N. 1996. Policy analysis, peer reviewed: Evaluation of bioaccumulation factors in regulating metals. *Environ. Sci. Technol.* 30(10):448A–452A.

CIM. 2010. CIM definitions standards for mineral resources and mineral reserves: Standing Committee on Reserve Definitions Adopted by CIM Council on November 27, 2010. http://web.cim.org/UserFiles/File/CIM_DEFINITON_STANDARDS_Nov_2010.pdf. Accessed 9/11/13.

Dowd, P.A. 2002. The assessment and analysis of financial, technical, and environmental risk in mineral resource evaluation. In *Deposit and Geoenvironmental Models for Resource Exploitation and Environmental Security*. Edited by A.G. Fabbri, G. Gaal, and R.B. McCammon. Netherlands: Kluwer. pp. 187–221.

ECHA (European Chemicals Agency). 2008. Appendix R.7.13-2: Environmental risk assessment for metals and metal compounds. In *Guidance on Information Requirements and Chemical Safety Assessment*. Guidance for the Implementation of REACH report. http://echa.europa.eu/documents/10162/13632/information_requirements_r7_13_2_en.pdf. Accessed 9/11/13.

Environment Canada. 2004. *National Air Pollution Surveillance Network Quality Assurance and Quality Control Guidelines*. AAQD 2004-1. Ottawa, ON: Environment Technology Centre.

Environment Canada. 2005. *Guidance for Fish Tissue Analysis for Mercury Using Non-Lethal Methods for the Metal Mining Environmental Effects Monitoring Program*. Report. www.ec.gc.ca/eem/pdf_publications/English/mm_fish_tissue.pdf. Accessed 9/11/13.

Environment Canada. 2012. *Metal Mining Technical Guidance for Environmental Effects Monitoring*. Report. www.ec.gc.ca/Publications/D175537B-24E3-46E8-9BB4-C3B0D0DA806D%5CCOM-1434---Tec-Guide-for-Metal-Mining-Env-Effects-Monitoring_En_02.pdf. Accessed on 9/11/13.

European Commission. 2012, Summary of EU legislation on mining waste. Directive 2006/21/EC of the European Parliament and of the Council on the Management of Waste from the extractive Industries. http://ec.europa.eu/environment/waste/mining/legis.htm and http://europa.eu/legislation_summaries/environment/waste_management/l28134_en.htm. Accessed 9/11/13.

European Communities. 2003 *Common Implementation Strategy for the Water Framework Directive (2000/60/EC)*. Guidance Document No 7: Monitoring Under the Water Framework. www.eurogeologists.de/images/content/panels_of_experts/hydrogeology/9E9DFd01.pdf. Accessed 9/11/13.

Failing, L., Gregory, R., and Harstone, R.M. 2007. Integrating science and local knowledge in environmental risk management: A decision-focused approach. *Ecolog. Econ.* 64:47–60.

Fairbrother, A., Wenstel, R., Sappington, K., and Wood, W. 2007. Framework for metals risk assessment. *Ecotoxicol. Environ. Safety* 68:145–227.

Finlayson C.M. 1996. Framework for designing a monitoring programme. In *Monitoring Mediterranean Wetlands: A Methodological Guide*. Edited by P. Tomas Vives. Slimbridge, United Kingdom; and ICN, Lisbon: Wetlands International. pp. 25–34.

Ford, K.L. 2004. *Risk Management Criteria for Metals at BLM Mining Sites*. Technical Note 390 (revised) BLM/RS/ST-97/001+1703. Denver, CO: U.S. Bureau of Land Management, National Science and Technology Center. www.blm.gov/nstc/library/pdf/TN390v04.pdf. Accessed 9/11/13.

Goudarzi, G.H. 1984. *Guide to Preparation of Mineral Survey Reports on Public Lands*. U.S. Geological Survey, Open-File Report 84-787. http://pubs.usgs.gov/of/1984/0787/report.pdf. Accessed 9/11/13.

GRI (Global Reporting Initiative). 2010. *Sustainability Reporting Guidelines and Mining and Metals Sector Supplement*. Global Reporting Report. https://www.globalreporting.org/reporting/sector-guidance/mining-andmetals/Pages/default.aspx. Accessed 9/11/13.

ICMM (International Council on Mining and Metals). 2006. *HERAG: Health Risk Assessment Guidance for Metals*. ICMM report. www.icmm.com/page/1213/health-risk-assessment-guidance-for-metals-herag. Accessed 9/11/13.

ICMM (International Council on Mining and Metals). 2007. *MERAG: Metals Environmental Risk Assessment Guidance*. www.icmm.com/page/1185/metals-environmental-risk-assessment-guidance-merag. Accessed 9/11/13.

ICMM (International Council on Mining and Metals). 2008. *A Sustained Commitment to Improved Industry Performance*. London: ICMM. www.corporate-citizenship.com/wp-content/uploads/Mining.pdf. Accessed 9/11/13.

ICMM (International Council on Mining and Metals). 2013. Sustainable development framework: 10 principles. London: ICMM. www.icmm.com/our-work/sustainable-development-framework/10-principles. Accessed 9/26/13.

INAP (International Network for Acid Prevention). 2009. *Global Acid Rock Drainage Guide* (GARD Guide). www.gardguide.com. Accessed 9/11/13.

ISO (International Organization for Standardization) 31000:2009. *Risk Management—Principles and Guidelines*. www.iso.org/iso/home/standards/iso31000.htm. Accessed 12/16/13.

Kaplan, S., and Garrick, B.J. 1981. On the quantitative definition of risk. *Risk Anal.* 1:11–27. http://josiah.berkeley.edu/2007Fall/NE275/CourseReader/3.pdf. Accessed 9/11/13.

Kuipers, J.R., Maest, A.S., MacHardy, K.A., and Lawson, G. 2006. Comparison of predicted and actual water quality at hardrock mines—The reliability of predictions in Environmental Impact Statements. Butte, MT: Kuipers and Associates; Boulder, CO: Buka Environmental. http://pebblescience.org/pdfs/ComparisonsReportFinal.pdf. Accessed 9/11/13.

Lacy, W. 2010. *An Introduction to Geology and Hard Rock Mining.* Westminster, CO: Rocky Mountain Mineral Law Foundation. www.rmmlf.org/scitech/lacy/lacy.htm. Accessed 9/11/13.

McLemore, V.T., Russell, C.C., Smith, K.S., and the Sampling and Monitoring Committee of the Acid Drainage Technology Initiative–Metal Mining Sector. 2004. Sampling and monitoring for closure. SME Preprint No. 04-62. Littleton, CO: SME. www.onemine.org/view/?d=BB60F953960F2274EBE26EC2304AA37CA5BD2E24D37573C249E8B807896ED8E038402. Accessed 9/11/13.

McLemore, V.T., Smith, K.S., Russell, C.C., and the Sampling and Monitoring Committee of the Acid Drainage Technology Initiative–Metal Mining Sector. 2007. Sampling and monitoring for closure. In *Understanding and Responding to Hazardous Substances at Mine Sites in the Western United States.* Edited by J.V. DeGraff. Reviews in Engineering Geology, Vol. 17. Boulder, CO: Geological Society of America. pp. 171–180.

McLemore, V.T., Russell, C.C., and Smith, K.S. 2009. ADTI-MMS sampling and monitoring for the mine-life cycle: Securing the Future and 8th ICARD, June 23–26, Skelleftea, Sweden, conference proceedings. www.proceedings-stfandicard-2009.com/pdfer/Virginia_McLemore_B4_T3_ADTI-MMS-Sampling-and-Monitoring-for-the-Mine-life-Cycle.pdf. Accessed 2/8/10.

Ministry of Environment. 2012. *Water and Air Baseline Monitoring Guidance Document for Mine Proponents and Operator*: British Columbia report. www.env.gov.bc.ca/epd/industrial/mining/pdf/water_air_baseline_monitoring.pdf. Accessed 9/11/13.

Miranda, M., and Sauer, A. 2010. Mine the gap: Connecting water risks and disclosure in the mining sector. WRI Working Paper. Washington, DC: World Resources Institute. http://pdf.wri.org/working_papers/mine_the_gap.pdf. Accessed 9/11/13.

Paterson, Grant and Watson Limited and Geomatics International, Inc. 1994. *Application of Remote Sensing and Geophysics to the Detection and Monitoring of Acid Mine Drainage*: MEND Project 4.6.3. www.mend-nedem.org/reports/files/4.6.3.pdf. Accessed 9/11/13.

Ramsar Convention. 1996. Resolution VI.1. In *Proceedings of the Sixth Meeting of the Conference of the Contracting Parties of the Convention on Wetlands, Resolutions and Recommendations*, Ramsar Convention Bureau, Gland, Switzerland.

Rouse, W.B. 2009. *Handbook of Systems Engineering and Management*. New York: John Wiley and Sons.

Russell, C. 2006. Mine drainage challenges from A to Z in the United States. In *Proceedings, 7th International Conference on Acid Rock Drainage (ICARD 7).* Lexington, KY: American Society of Mining and Reclamation. pp. 1785–1802, www.imwa.info/docs/imwa_2006/1785-Russell-CO.pdf. Accessed 9/11/13.

SEC (U.S. Securities and Exchange Commission). 2013. Industry Guide 7: Description of Property by Issues Engaged or To Be Engaged in Significant Mining Operations. Washington, DC: SEC. http://web.cim.org/standards/documents/Block474_Doc32.pdf. Accessed 9/11/13.

SENES Consultants Limited. 1994. *Handbook for Waste Rock Sampling Techniques*. MEND Project 4.5.1-2. www.mend-nedem.org/reports/files/4.5.1-2.pdf. Accessed 9/11/13.

SENES Consultants Limited, Golder Associé Ltée, and Laval University. 1994. *Review of Waste Rock Sampling Techniques*. MEND Project 4.5.1-1. www.mend-nedem.org/reports/files/4.5.1-1.pdf. Accessed 9/11/13.

SME (Society for Mining, Metallurgy and Exploration). 2007. The SME Guide for Reporting Exploration Results, Mineral Resources, and Mineral Reserves (The 2007 SME Guide). Littleton, CO: SME. www.crirsco.com/usa_sme_guide_2007.pdf. Accessed 9/11/13.

Smith, K.S. 2007. Strategies to predict metal mobility in surficial mining environments, Ch. 3. In *Understanding and Responding to Hazardous Substances at Mine Sites in the Western United States*. Edited by J.V. DeGraff. Reviews in Engineering Geology, Vol. 17. Boulder, CO: Geological Society of America. pp. 25–45.

Smith, K.S., and Huyck, H.L.O. 1999. An overview of the abundance, relative mobility, bioavailability, and human toxicity of metals. In *The Environmental Geochemistry of Mineral Deposits—Part A: Processes, Techniques, and Health Issues*. Reviews in Economic Geology, Vol. 6A. Littleton, CO: Society of Economic Geologists. pp. 29–70.

Suter II., G.W. 1993. *Ecological Risk Assessment*. Boca Raton, FL: Lewis Publishers.

Terrestrial and Aquatic Environmental Managers and SENES Consultants Limited. 1997. Appendix A: Technical summary note. In *Guideline Document for Monitoring Acid Mine Drainage*. MEND Project 4.5.4. www.mend-nedem.org/reports/files/4.5.4%20Appendix.pdf. Accessed 9/11/13.

USEPA (U.S. Environmental Protection Agency). 2001a. *Abandoned Mine Site Characterization and Cleanup Handbook*. EPA910-B-00-001. www.epa.gov/superfund/policy/remedy/pdfs/amscch.pdf. Accessed 9/11/13.

USEPA (U.S. Environmental Protection Agency). 2001b. *EPA Requirements for Quality Assurance Project Plans* (EPA QA/R-5): U.S. Environmental Protection Agency, EPA/240/B-01/003. www.epa.gov/quality1/qs-docs/r5-final.pdf. Accessed 12/20/13.

USEPA (U.S. Environmental Protection Agency). 2001c. *EPA Requirements for Quality Management Plans* (EPA QA/R-2): U.S. Environmental Protection Agency, www.epa.gov/quality/qs-docs/r2-final.pdf. Accessed 12/20/13.

USEPA (U.S. Environmental Protection Agency). 2002a. *Guidance on Choosing a Sampling Design for Environmental Data Collection for Use in Developing a Quality Assurance Project Plan* (EPA QA/G-5S): U.S. Environmental Protection Agency, www.epa.gov/QUALITY/qs-docs/g5s-final.pdf. Accessed 9/11/13.

USEPA (U.S. Environmental Protection Agency). 2002b. *EPA Guidance for Quality Assurance Project Plans* (EPA QA/G-5): U.S. Environmental Protection Agency, www.epa.gov/quality/qs-docs/g5-final.pdf. Accessed 12/20/13.

USEPA (U.S. Environmental Protection Agency). 2003. *EPA and Hardrock Mining: A Source Book for Industry in the Northwest and Alaska*. EPS report. http://yosemite.epa.gov/R10/WATER.NSF/Sole+Source+Aquifers/hardrockmining. Accessed 9/11/13.

USEPA (U.S. Environmental Protection Agency). 2006. *Guidance on Systematic Planning Using the Data Quality Objectives Process*. EPA QA/G-4. Office of Environmental Information EPA/240/B -06/001. www.epa.gov/QUALITY/qs-docs/g4-final.pdf. Accessed 9/11/13.

USEPA (U.S. Environmental Protection Agency). 2007. *Framework for Metals Risk Assessment*. EPA 120/R-07/001. www.epa.gov/raf/metalsframework/pdfs/metals-risk-assessment-final.pdf. Accessed 9/11/13.

USEPA (U.S. Environmental Protection Agency). 2011. Framework for Metals Risk Assessment Web page. Office of the Science Advisor. www.epa.gov/raf/metalsframework/index.htm. Accessed 12/20/13.

USEPA (U.S. Environmental Protection Agency). (forthcoming). Guidance Manual for Environmental Site Characterization in support of Environmental and Human Health Risk Assessment.

Waddell, N. 2000. *Internal Quality Assurance and Quality Control for AusRivAS in NSW*. Monitoring River Health Initiative Technical Report No. 2b, Commonwealth of Australia and NSW Environment Protection Authority, Canberra and Sydney. www.environment.gov.au/water/publications/environmental/rivers/nrhp/pubs/nsw-quality.pdf. Accessed 9/11/13.

CHAPTER 5

The Planning Process

Carol C. Russell, Kathleen S. Smith, and Virginia T. McLemore

5.1 INTRODUCTION

The planning process is a structured, systematic approach for obtaining data and providing information to make decisions. The basis of the planning process is the definition of data quality objectives (DQOs) that result in the sampling and monitoring plan, described in Chapter 6. The DQO process is used to develop performance and acceptance criteria that clarify study objectives, define the appropriate type of data, and specify tolerable levels of potential decision errors that will be used as the basis for establishing the quality and quantity of data needed to support decisions (USEPA 2006). By selecting a representative part of a population (sampling) and observing or checking it over a period of time (monitoring), these data can be analyzed to provide information to guide decision making. Sampling strategies should be site specific and consider characteristics such as climate, geology, proximity of viable resources, and unique characteristics of the site. The primary purpose for using a quality planning process is to ensure that data is collected that is

- Needed to make decisions,
- Can be used to make decisions, and
- Will be used to make decisions.

In designing a monitoring system, the first four steps shown in Figure 4.1 in Chapter 4 constitute the planning process portion of the sampling and monitoring program. They include

- Define the questions and objectives (who is asking the question?);
- Develop the site conceptual model;
- Evaluate the risks, consequences, and costs of not sampling and monitoring; and
- Decide what is at risk if the information is incorrect or not available (how sufficient and complete the data should be to answer the questions).

Examples of sampling and monitoring programs are provided in the Types of Monitoring Programs sidebar and in Appendix 4.

5.2 DEFINE QUESTIONS AND OBJECTIVES

Defining the objectives may seem simple, but actually development of the objectives is the most important and sometimes the most difficult component of the sampling and monitoring program. What questions are being asked by decision makers and what are their objectives? Who are these decision makers, what decisions are they making, and what information do they need to make better decisions? High-quality, credible scientific information is undeniably the best foundation for environmental decision making. Quite often, the questions are associated with environmental

TYPES OF MONITORING PROGRAMS

Virginia T. McLemore, New Mexico Bureau of Geology and Mineral Resources

Different types of monitoring programs need to be considered: compliance monitoring, trend monitoring, impact monitoring, and survey monitoring. *Compliance monitoring* is used to determine if permit objectives are being met. The objectives are site specific, but the data obtained can be used for other purposes such as modeling or prediction. *Trend monitoring* is used to detect changes over time, typically 10 years or more, to ensure that the trends are detected, and the program should be consistent throughout the entire sampling period (i.e., frequency, location, time of day, sampling and analytical techniques employed). An example of trend monitoring would be measurements of water quality in monitoring wells surrounding a tailings facility. *Impact assessment monitoring* is used to identify, characterize, and measure the effects of a mine facility or event on the environment, such as water quality. It is best that both test and control sites are included. A baseline assessment to determine premining conditions would be an example of impact assessment monitoring. *Survey or inventory monitoring* is used to characterize the site conditions at a specific location and time. A watershed approach is generally the most effective.

compliance. Consider the questions more broadly to incorporate other sustainability concerns including the economic and social aspects of the mining and processing operation in addition to environmental concerns. Potential questions include the following:

- What is the current status of the environment?
- Are there any aspects of the environment of particular value or concern?
- What are the problems?
- What is the nature and extent of the problem?
- How effective is the solution?
- How much will the mitigation cost, and do the stakeholders agree?

Mining and processing operations are as unique as the challenges associated with them. However, over time a few key lessons have been learned about answering questions regarding potential impacts at mining sites. First, knowing the environmental goals throughout the mine life cycle is essential. Second, the use of a conceptual model helps to evaluate risks by looking at the sources of potential contamination, the entity to be protected (receptor), such as human health or endangered species, and the pathway between the two. Third, defining risks at the mine site helps to define questions: from the risk of a landslide to the costs of Superfund listing. Fourth, the types of information needed to answer questions often revolve around water where using a conceptual model for the preparation of a water balance calculation throughout the mine life cycle is of paramount importance. The water balance will help to answer the inevitable question of *Will there be a discharge of mine dewatering or process wastewater during operations and/or closure, and will the discharge meet water quality standards?* Fifth, the question of mitigation measures revolves around identifying the measures to be taken at what times, in what manner, and by whom.

A corporate strategy to address environmental risks is similar to that needed for mine planning for maintaining ore grade, and they should be performed concurrently. Decision making based on scientific/engineering information assists in the evaluation of the operation's viability and results in time and cost savings. The need for additional data collection or the costly mistake of reanalysis at lower detection levels can easily be avoided with proper planning. Regulatory agencies would also be glad to have sufficient data to evaluate the potential environmental impacts in an efficient and effective manner.

5.2.1 Determine Data Quality Objectives

For decision problems, the DQOs are typically expressed as tolerable limits on the probability or chance (risk) of the collected data leading one to making an erroneous decision. The concept of intended use of the data is used to set the context for planning activities and focus the attention of the planning team. For example, one company's objective was to never exceed the water quality standard for copper in their water discharge. The intended use of the data was to assure the president of the company that they were in legal compliance. In order to ensure that 100% of the time that they would not discharge in violation, the average discharge concentrations needed to be reduced by more than half and a new treatment technology would need to be developed. From a cost-effective perspective, the company may have been better off to have the objective be in compliance for 98% of the time with a ±2% uncertainty. For estimation problems, the DQOs are typically expressed in terms of acceptable uncertainty (e.g., width of an uncertainty band or interval) associated with a point estimate at a desired level of statistical confidence.

5.2.2 Define the Confidence Level

A data collection design specifies the type, number, location, and physical quantity of samples and data, as well as the quality assurance/quality control (QA/QC) activities that will ensure that sampling design and measurement errors are managed sufficiently to meet the performance or acceptance criteria specified in the DQOs. As discussed previously, prior to beginning a sampling and monitoring program, it is necessary to define the reasons for sampling, the questions to be answered, and the desired degree of confidence in the answer(s). One might ask

- How much data are needed?
- What quality is sufficient?

The desired degree of confidence should be defined, as much as possible, at the inception of an investigation. A low level of certainty in the data can lead to erroneous interpretations and flawed decisions, but a higher level of certainty generally requires more sampling and higher costs. The degree of QA/QC should be in proportion to the degree of confidence that is required for the decisions to be made. Many situations exist where high-confidence data are necessary. For example, if expensive decisions are going to be made or if lives are at stake based on the outcome of the investigation, then high-confidence data are warranted. Other situations in which high-confidence data are likely desirable are for risk assessments or when data will be used in litigation. In one instance, a sampling plan was developed for the sole purpose of getting two people into the field together to allow them to talk. The quality of the data was immaterial. Table 5.1 provides an overview of the sources of uncertainty and variability.

It is necessary to select both the type of confidence and the level of confidence desired in the investigation. Types of confidence include statistical confidence and professional confidence. Confidence level is a statistical concept, which is based on probability, discussed further in Chapter 7.

Environmental samples are generally heterogeneous and subject to variability. This variability produces a pattern, which is called a *distribution*. The distribution can be described mathematically and graphically. The best-known theoretical distribution is the bell-shaped normal distribution. Most environmental and chemical data more closely follow a log-normal distribution, where the logarithms of data approximate a bell-shaped curve. A number of statistical tests and descriptions can be derived for normal (or log-normal) distributions. For example, the confidence interval is

TABLE 5.1 Variability versus uncertainty in risk assessment

For uncertainty analysis, it is helpful to verify whether the parameter under consideration is characterized by variability, uncertainty, or both. This verification is based on the definition of *variability* and *uncertainty* and the scope of the assessment. An overview of sources of uncertainty and variability in risk assessments of metals is given in the table below.

	Variability	Uncertainty
	• Not reducible through further research • Probability distribution represents real variations actually occurring	• Reducible through further research • True value is somewhere in the probability distribution
Exposure	• Temporal variability of release • Spatial variability of soil/sediment/ aquatic characteristics and background concentrations (regional assessment)	• Uncertainty of models • Uncertainty of physiochemical properties, partition coefficients, removal rates • Spatial variability of soil/sediment/ aquatic characteristics and background concentrations (local assessment)
Effects	• Intra- and inter-species sensitivity • Inter- and intra-laboratory variability • Endpoint differences • Spatial (and temporal) variability of physiochemical characteristics determining bioavailability	• Probability distribution uncertainty (e.g., threshold versus nonthreshold distribution) • Diversity representativeness • Sampling uncertainty • Lab to field extrapolation

Source: Adapted from MERAG 2007.

a range of values that has a specified probability of containing the population parameter being estimated. In the case of the mean, it is the range of values that trap the population mean a certain percentage of the time (Figure 5.1).

Confidence limits are the lower and upper boundaries (or values) of a confidence interval. The confidence level is the probability value associated with a confidence interval (i.e., the probability that the population value is contained within the interval) and is often expressed as a percentage (Figure 5.1). The confidence level is an important detail to consider when designing a sampling plan (USEPA 2006).

5.2.3 Risks If the Data Are Unavailable or Incorrect

There is an entire field of science dedicated to a methodology of inquiry that is appropriate for cases where "facts are uncertain; values are in dispute; stakes are high and decisions urgent" (Funtowicz and Ravetz 1991). It is primarily applied in the context of long-term issues where there is less available information than is desired by decision makers and stakeholders. Practically speaking, the most significant risk to a mining company is time, and therefore money lost, when data are unavailable or incorrect going into the environmental permitting process. The Tale of Two Mines sidebar demonstrates the importance of defining objectives and making sure one has the information needed. In this sidebar, characteristics on the mines have been obscured, but the issues and relative span of times between the two approvals are real.

5.3 DEVELOP SITE CONCEPTUAL MODELS

The most important and probably the simplest tool for evaluating the environmental impacts of a project is the development of a conceptual model. The creation of a conceptual model for premining (exploration), during the operations, closure, and postclosure is an important step in prediction of potential environmental impacts at mining projects. The conceptual model should

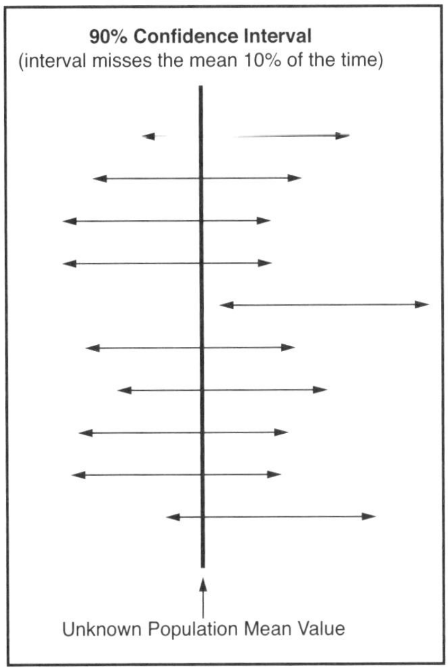

FIGURE 5.1 Concept of a 90% confidence interval

TALE OF TWO MINES

Carol C. Russell, U.S. Environmental Protection Agency

Two mining environmental impact statements (EISs) were simultaneously undergoing review by the same personnel within the same regulatory agencies. Mine One had little appropriate data to answer questions the regulatory agencies were asking: Would discharges from the mine meet water quality standards? Would wetlands and endangered species be impacted? Would the proposed water management system be adequate? Data were ±20 years out of date, with detection levels above standards, utilizing outmoded analytical methods, including sampling locations that were not applicable to the new mine design.

In comparison, Mine Two was proactive. They identified constituents of concern, implemented a sampling and monitoring program in cooperation with a local stakeholders group, and designed and implemented all necessary mitigation measures.

Mine Two got their permit in approximately a year. Mine One, after two EISs with adverse ratings and several lawsuits, has yet to receive their permit after more than 20 years.

be used to enable experts from all disciplines to communicate effectively with one another, resolve issues concerning the site, and facilitate the decision-making process. The conceptual model is used to integrate all site information and to determine whether information, including data, is missing (data gaps) and whether additional data need to be collected at the site. A conceptual model is considered to be a dynamic tool that will allow for hypothesis testing of the concepts. Furthermore, the model is used to facilitate the selection of mitigation alternatives and to evaluate the effectiveness of remedial actions in reducing the exposure of environmental receptors to contaminants.

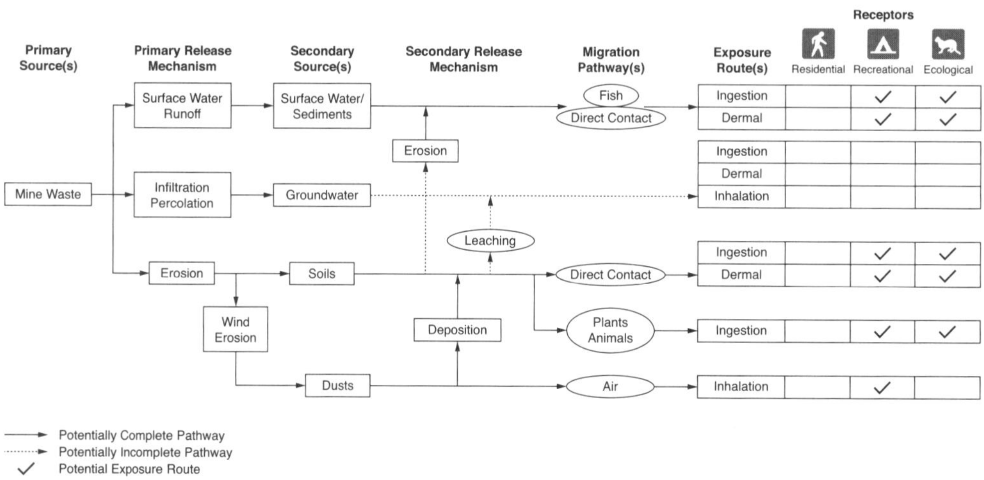

Source: Ford 2004.

FIGURE 5.2 Mine water conceptual model for human and ecological risk

5.3.1 Review Existing Data

The basis of a conceptual model is currently available data. Because new data collection can be expensive and time-consuming, it is important to conduct a literature and company review of all existing data and information concerning the site. This is a step that is often overlooked or cut short. These data are not merely limited to the mining company and consultants but can include data from local, state, and federal agencies; environmental groups; universities; and other entities. It is helpful to consider how future data collection can make use of or complement existing data, and how existing data can be used to help determine baseline conditions. Local land-use plans need to be investigated and nearby sensitive land-use areas identified (e.g., conservation areas, critical habitats, wetlands, national parks, cultural sites, recreational areas). It is also important to collect information that provides the *social* setting of the area (e.g., demographics; land ownership; proximity to structures, roads, and waterways; cultural and community features). This existing information can be used to build the initial conceptual model.

5.3.2 Purpose and Use of the Conceptual Model

By definition, the conceptual model will be an abstraction that aims to identify those key processes that affect contaminant transport, behavior, and risk. The purpose of the conceptual model is to illustrate and describe a basic understanding of potential sources, media pathways, and possible receptors, as shown in Figure 5.2.

The ASTM standard for development of a conceptual model (E 1689-95) advises the development of (1) a flow diagram that illustrates potential sources, transport mechanisms, exposure pathways and receptors; (2) a schematic block diagram that depicts these site model components; and (3) a map of surface units and other relevant features (ASTM 2008).

5.3.3 Elements of Conceptual Models

Source. At a mine site, the source may be mine and processing waste, such as a tailings impoundment, an area of contaminated soil or sediment from which metals may be released, or the actual mine pit or underground workings.

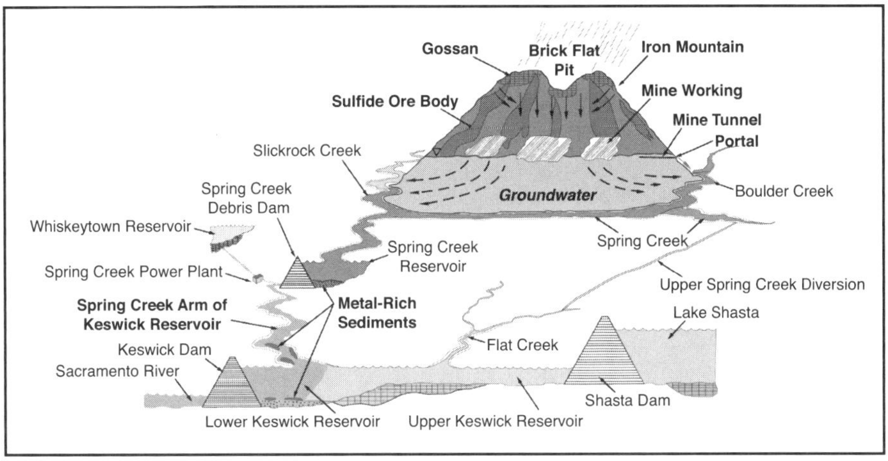

Source: INAP 2009.

FIGURE 5.3 Conceptual model showing metal and acid source regions at Iron Mountain and downstream transport pathways to the Sacramento River, California

Exposure pathways. The classic pathways for exposure at a mine site are transport via air (e.g., fugitive dust), groundwater (e.g., contaminated plumes), or surface water (Figure 5.2; e.g., runoff). There are situations at mine sites wherein the pathway is a flow of solid (e.g., waste-rock pile slump) or semisolid (e.g., tailings released from an impoundment) waste materials released from a waste storage facility. In addition to these types of exposure pathways that take the hazard from the source to the receptor, the receptor may actually come to the source for exposure (direct contact). Examples of this include migratory wildfowl landing on contaminated ponds or children playing in contaminated soils; both situations have been observed at historical mine sites.

Receptors. Historically, the primary receptors of concern at mine sites have been humans. This includes people living or working at the mine site, visitors to the site, and people living nearby the site. Additional receptors or endpoints may drive the project manager's response, including aquatic species (e.g., fish and invertebrates), terrestrial wildlife (e.g., invertebrates, birds, and mammals), and floral populations. Receptors may not be present at the time the sampling takes place.

Figure 5.3, a graphic of the conceptual flow pathways at the Iron Mountain mine site, depicts site model components and a map of surface units and other relevant features within the assessment boundary. The most important sources are mine waste, ore, process waste, mine workings, and ore extraction and processing. Water is a primary environmental pathway for constituents released from these sources. Transport occurs by groundwater, surface water, or infiltration through the vadose zone. Because water is a primary pathway, aquatic resources are generally the receptor of most interest.

Uncertainties associated with the conceptual site model need to be identified clearly so that efforts can be taken to reduce these uncertainties to acceptable levels. Early versions of the model, which are usually based on limited or incomplete information, will identify and emphasize the uncertainties that should be addressed.

5.3.4 Developing a Water Balance Conceptual Model

Water balance models are mathematical computation procedures to describe and quantify the spatial and temporal distribution of the water in the forms of precipitation, evapotranspiration,

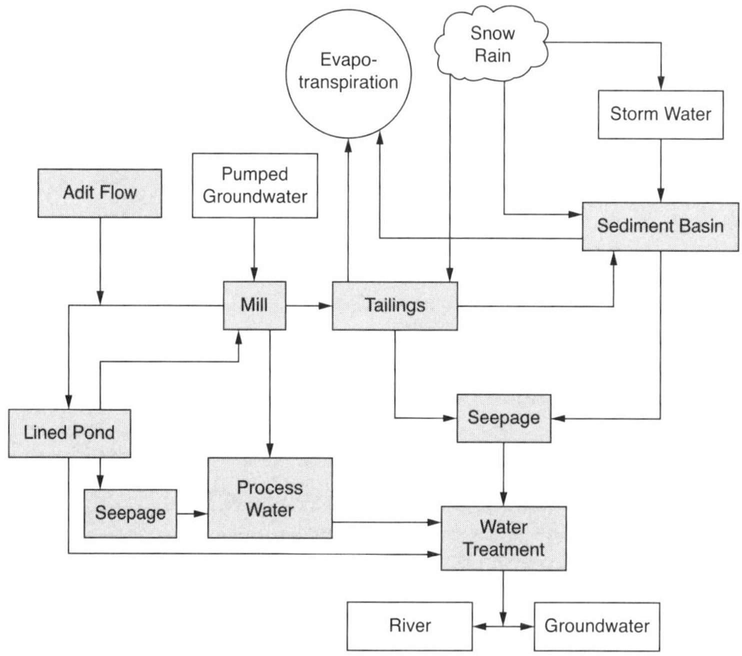

FIGURE 5.4 Flowchart describing water balance as a conceptual model

infiltration, water storage, treatment, seepage, runoff, and discharge (Figure 5.4). The water balance should include surface water and groundwater. It should be robust and provide information on the probable scenarios that might be encountered at the mine site over the full mine cycle.

Development of a site water balance can be facilitated by drafting a conceptual model. The hydrological parameters (e.g., flow or discharge measurements, groundwater flow parameters) and climatological data (e.g., precipitation, temperature, evapotranspiration; see Figure 5.5) information are the key components of a water balance. Mass balance of water sources (inputs) and losses (outputs) need to be identified in the sampling and monitoring plan to quantify each flow pathway (Figures 5.3 and 5.4). Tracers are a useful technique to determine gain or loss of water quantity in streams (see Appendix 3). Seasonal effects are often an important aspect of conceptual models of water balance to examine and quantify. The U.S. Environmental Protection Agency found that a common thread among active Superfund mine sites is that they did not have an accurate water balance (Russell 2006). The Methods for Analyzing Cyanide Needed sidebar provides an example of the importance of water balance.

5.4 CONCEPTUAL MODELS FOR NUMERICAL MODELING

Data availability is one source of uncertainty in the development of models for decision support. To minimize uncertainty, development of a reasonable conceptual model helps to identify data gaps that need to be to populated before using numerical models. Although there are other sources of uncertainty, models should be selected (simple vs. complex) in part based on the data available to support their use. For example, a simple analysis using models like those described as screening procedures could be run quickly at low cost to begin to understand the issues (Mills et al. 1985). This understanding might suggest (perhaps through sensitivity analysis) that data should be collected on current land use, or that a limited monitoring program is warranted. Following acquisition of that data, a revised (perhaps more detailed) model could be developed. This

Courtesy of Virginia T. McLemore.

FIGURE 5.5 Measuring water infiltration using an infiltrometer, Questa rock piles, Taos County, New Mexico

METHODS FOR ANALYZING CYANIDE NEEDED—BEAL MOUNTAIN MINE

Carol C. Russell, U.S. Environmental Protection Agency

In 1999, the Montana Department of Environmental Quality and the U.S. Forest Service took over the management of water treatment for the bankrupt Beal Mountain mine near Butte, Montana. A trustee was appointed to oversee disbursement of funds, including the bond money for mine water process water treatment. Approximately 1 billion liters of heap-leach pad process solutions have been reduced to 50 million liters with the use of a biological treatment plant and a land application system (W. McCollough, personal communication, June 2003). Although the solutions met the state water-quality standards after treatment for pH and for weak-acid dissociable (WAD) cyanide, all plants within the land application area died within a short time. The operator determined, through greenhouse tests of the water, that the phytotoxicity was due to thiocyanates (B. Parker, personal communication, December 2002). During treatment for metals and pH, many of the breakdown compounds of cyanide, while generally less toxic than the original cyanide, are known to be toxic to aquatic organisms and can persist in the environment for significant periods of time. Some of these potentially toxic breakdown forms include cyanates, ammonia, nitrite, nitrate, and thiocyanates. These compounds, however, were not identified for analysis; only total cyanide or WAD analysis were requested, and thiocyanate is a known interference for the cyanide distillation methods used in the United States. For example, water samples from mining and processing sites where cyanide is used as a process chemical can have WAD and/or total cyanide concentrations that are quite low or undetected, yet when the same samples are analyzed specifically for cyanates and thiocyanates, they show tens of milligrams per liter or more of these compounds (Russell 2006). The lesson learned in this situation is that when treating for acid rock drainage, new compounds can be formed that need to be evaluated for toxicity.

strategy for data-poor situations makes efficient use of resources and targets the effort toward information and models that will reduce the uncertainty as the analysis proceeds. On the other end of the spectrum is the PHREEQC aqueous geochemistry model (Parkhurst and Appelo 2013), a very complex and useful model when sufficient data are available. Frequently, physical parameters are unavailable, making most geochemical models useless. The use of complex mechanistic models is warranted if it helps promote understanding of complex systems and as long as

WATER QUALITY DATA NEEDS FOR TMDLs

Flow data. Critical to the process of calibrating and verifying models are flow data from sources and various locations in the receiving water. Flow data are generally high in quality if gathered as part of unidirectional stream surveys. The U.S. Geological Survey (USGS) is generally considered to be the reliable source for long-term, high-quality data sets.

Ambient water quality data. A number of federal agencies, state agencies, regional organizations, and research groups collect surface water quality data. Many of these data are retrievable over the Internet, particularly data from USGS and the U.S. Environmental Protection Agency (USEPA). Although there is no universal repository for all surface water quality data, the STORET database is the most comprehensive. Because methods of collection and analysis may vary, there is a need for quality assurance and quality control of these data.

Land use data. All states should have access to a series of land use records and projections. For ease of use, the land use data sets should be made available as geographic information system coverages. USEPA has provided default coverages as a component of its BASINS model. For total maximum daily load (TMDL) purposes, land use data are required for the time period over which water quality data are available in order to calibrate and validate models. Projected land use data are needed for predicting future scenarios. The overall quality of these land use data will vary, often as a function of the level of ground-truthing that was done or the accuracy of the predictions for future land use changes.

Point source data. Model inputs may include measured values of pollutant loading from point sources (e.g., based on information reported on National Pollutant Discharge Elimination System discharge monitoring reports submitted by permitted facilities). Other possible data sources include results from periodic compliance inspection and wasteload allocation studies or data collected as part of field surveys done in support of the TMDL. Such data are generally available and more reliable than data from point sources. This is partly because during high-flow, high-rainfall events, monitoring is only infrequently conducted. Mining processing and storm-water discharges are usually considered to be point sources.

Atmospheric deposition. Data on pollutant loadings from atmospheric deposition have been compiled by the National Atmospheric Deposition Program. Atmospheric nitrate and nitrate from explosives such as ammonium nitrate may be comingled.

Source: Adapted from NRC 2001.

uncertainties in the results are reported and incorporated into decision making. However, there may be a tendency to use complex mechanistic models to conduct water quality assessments in situations with little useful water quality data and/or involving major remediation expenditures or legal actions. In these situations, there is usually a common belief that the realism in the model can compensate for a lack of data, and the complexity of the model gives the impression of credibility. However, given that uncertainty in models is likely to be exacerbated by a lack of data, the recommended strategy is to begin with a simple modeling study and iteratively expand the analysis as needs and new information dictate.

5.4.1 Total Maximum Daily Load Models

Water quality models are used for watershed water quality modeling of contaminant loads and to evaluate the potential for violation of water quality standards or criteria. The data required for total maximum daily load (TMDL) model development is a function of the water criterion or standard and its location and the analytical procedures used to relate the stressors to the criterion similar to the needs of a risk-based conceptual model. Data needs can include hydrology (streamflow, precipitation), ambient water quality measures, and land use and elevation. The U.S. Geological Survey has provided excellent methods specifically for mining TMDLs as expanded further in the references (Walton-Day et al. 2002, 2012). TMDL development will also likely require data on point/nonpoint sources and pollutant loads, atmospheric deposition, the effectiveness of current best management practices, and legacy/upstream pollutant sources. Because the amount of available data varies with each site, there is no absolute minimum data requirement that can be universally set for TMDL development.

The data and information required for TMDL modeling must reflect the parameters that affect attainability of water quality standards. Many of the models used today have extremely large data requirements that must be addressed prior to project development. To ensure one has the proper data inputs, the modeler should be involved in the planning process for data collection. See the Water Quality Data Needs for TMDLs sidebar for a discussion of such data inputs.

5.5 SUMMARY

A well-designed and implemented program has the potential to deliver real value to mining companies and the public, both in terms of better environmental and financial decision making, and in the effectiveness and public credibility of the project for stockholders and stakeholders.

5.6 REFERENCES

ASTM E 1689-95 2008. *Standard Guide for Developing Conceptual Site Models for Contaminated Sites.* West Conshohocken, PA: ASTM International. www.astm.org/Standards/E1689.htm. Accessed 9/11/13.

Ford, K.L. 2004. *Risk Management Criteria for Metals at BLM Mining Sites.* Technical Note 390 (revised) BLM/RS/ST-97/001+1703. Denver, CO: U.S. Bureau of Land Management, National Science and Technology Center. www.blm.gov/nstc/library/pdf/TN390v04.pdf. Accessed 9/11/13.

Funtowicz, S., and Ravetz, J.R. 1991. A new scientific methodology for global environmental issues. In *Ecological Economics: The Science and Management of Sustainability.* Edited by R. Costanza. New York: Columbia University Press. pp. 137–152.

INAP (International Network for Acid Prevention). 2009. *Global Acid Rock Drainage Guide* (GARD Guide) www.gardguide.com. Accessed 9/11/13.

MERAG (Metals Environmental Risk Assessment Guidance). 2007. *Uncertainty Analysis.* MERAG Fact Sheet 07. www.icmm.com/document/259. Accessed 9/11/13.

Mills, W.B., Porcella, D.B., Ungs, M.J., Gherini, S.A., Summers, K.V., Mok, L., Rupp, G.L., Bowie, G.L., and Haith, D.A. 1985. *Water Quality Assessment: A Screening Procedure for Toxic and Conventional Pollutants in Surface and Groundwater* (Revised 1985). Part 1 EPA/600/6-85/002a. U.S. Environmental Protection Agency, Environmental Research Laboratory. http://yosemite.epa.gov/water/owrccatalog.nsf/e673c95b11602f2385256ae1007279fe/1d843503b32cc14485256b060072591c!OpenDocument. Accessed 9/11/13.

NRC (National Research Council). 2001. *Assessing the TMDL Approach to Water Quality Management.* Washington, DC: National Academies Press.

Parkhurst, D.L., and Appelo, C.A.J. 2013. *Description of Input and Examples for PHREEQC Version 3—A Computer Program for Speciation, Batch-Reaction, One-Dimensional Transport, and Inverse Geochemical Calculations.* U.S. Geological Survey, Techniques and Methods, Book 6. http://pubs.usgs.gov/tm/06/a43/pdf/tm6-A43.pdf. Accessed 9/11/13.

Russell, C. 2006. Mine drainage challenges from A to Z in the United States. In *Proceedings, 7th International Conference on Acid Rock Drainage (ICARD 7).* Lexington, KY: American Society of Mining and Reclamation. pp. 1785–1802, www.imwa.info/docs/imwa_2006/1785-Russell-CO.pdf. Accessed 9/11/13.

USEPA (U.S. Environmental Protection Agency). 2006. *Guidance on Systematic Planning Using the Data Quality Objectives Process EPA QA/G-4.* EPA/240/B -06/001. Washington, DC: Office of Environmental Information. www.epa.gov/QUALITY/qs-docs/g4-final.pdf. Accessed 9/11/13.

Walton-Day, K., Kimball, B.A., and Runkel, R.I. 2002. The use of mass-loading studies and solute transport modeling to assist in the development of TMDLs for streams affected by mine drainage. In Workshop Notebook, Colorado Department of Health and Environment 10th National Nonpoint Source Monitoring Workshop. Colorado Department of Public Health and Environment.

Walton-Day, K., Runkel, R.L., and Kimball, B.A. 2012. Using spatially detailed water-quality data and solute-transport modeling to support total maximum daily load development. *J. Am. Water Resour. Assn.* 48(5):949–969.

CHAPTER 6

Sampling and Monitoring Program Implementation

Carol C. Russell, Kathleen S. Smith, and Virginia T. McLemore

6.1 INTRODUCTION

Implementation of a sampling and monitoring program begins with a good SMP (sampling and monitoring plan). SMPs are written documents that outline procedures for obtaining data of adequate quality to ensure that the sample, analytical data, and scientific interpretations are sufficient to meet the goals and objectives of the program. The SMP (e.g., Table 6.1) should include the project management information/objectives, data acquisition (sampling methods, measurement and analysis procedures, data collection, or generation methods), data handling/analysis and data validation/usability. Several examples of sampling plans and a U.S. Environmental Protection Agency (USEPA) checklist on quality assurance (QA) document review can be found in Appendix 4. The field and analytical methods are often referred to as standard operating procedures (SOPs), many of which are briefly described in Appendix 3 (USEPA 2007). Another document used in sampling is the quality assurance project plan (QAPP), which describes the policy, organizational, and functional activities necessary to collect data that will stand up to legal and scientific

TABLE 6.1 Sampling and monitoring plan outline

Project Management	Project Organization/Responsibilities
	Problem Identification
	Data Quality Objectives
	Training Requirements
	Documentation
Data Acquisition and Management	Sampling Design and Methods of Collection
	Sample Handling and Custody
	Analytical Methods
	Quality Control Requirements
	Instrument/Equipment Testing, Inspection, and Maintenance
	Data Management
Data Validation	Validation and Verification Methods
	Data Review, Validation, and Verification
	Correspondence with Data Quality Objectives

Source: Adapted from USEPA 2013.

scrutiny (USEPA 2001). For our purposes, the distinctions between these documents will not be emphasized, but the need for documentation of the concepts described is the overriding issue. An example checklist is provided in Appendix 4.

6.2 SAMPLING AND MONITORING PLAN

The SMP is formal documentation of what will be measured, at what times, for what media, in what manner, and by whom based on the planning done as described in Chapter 5. At this stage, the questions and objectives have been defined, the conceptual model has been developed, the types of media and protocols have been chosen, the confidence levels with appropriate quality assurance and quality control (QA/QC; USEPA 2002a, 2002b) recognized, and the budget established. The following sections are suggested topics to consider in addition to the SMP outline. They will explore major concepts or headings with potential action items to fulfill each need.

6.2.1 Management, Leadership, and Organization

The project manager's responsibility is to the team and the program. He or she functions as the ultimate field manager to get useful data as planned in the first phases of the program. The manager is the one who ensures that the details have been addressed. For example, one key detail is that the monitoring locations are accurately identified, are safe for access, and that right of access has been secured. Permission for access to obtain samples is often facilitated by the offer to share results with the owner of the property being sampled. Other management tasks are QA/QC including chain-of-custody documentation, identification of project schedule, resources, milestones, and identification of any applicable safety requirements such as steel-toed boots or Mine Safety and Health Administration certification.

Leadership is also critical to provide team cohesiveness to meet the monitoring goals and to ensure that the sampling plan is implemented properly. The leader or project manager should be the one to officially sign off and date the SMP. Not only does the leader need to know the plan intimately but also must convey the purpose and all information to the team. One way to document the roles and responsibilities of the team is an organization chart with assigned duties. The sampling crew may be working under long and physically trying conditions. The leader is the one to observe the individuals and to call the end of the day when appropriate. The leader is also the one to facilitate positive working relationships among the team members. Coordination, communication pathways, and a process for resolving disputes in the SMP makes for a better team at the end of a trying day. The SMP should allow for unexpected delays, such as training of personnel, changes in personnel, ordering and shipping of materials, equipment malfunctions, and adverse weather conditions. In a similar fashion, the responsibility of each team member is to speak up if something is not going right.

Ultimately, the leader must have identified who the decision maker will be during each step of the SMP. For example, there is a decision maker who will be the user of the data and information developed as part of this program. Is that person the agency manager who determines if a permit is issued? Is it the new mine manager who will need to design and implement mitigation measures? Or is it the monitoring program manager who will provide information to the end user of the data? Development of the process for reporting results to the end users or decision makers, including the timing, format, and other expectations, should be anticipated before going in the field.

The leader should also have a communication strategy for proving information to stakeholders. Having a resident come up to the team and ask what you are doing requires forethought from the team member as to an appropriate response. Stakeholders are parties who may be affected by the results of the study and/or persons who may later use the data resulting from the data quality objective process (USEPA 2006). It is suggested that the manager develop a table listing the impacted organizations/stakeholders and the individuals who are representing those organizations. The table should also include how to contact them in order for the sampling team to know how to respond or to whom questions should be directed.

6.2.2 Team Qualifications and Training

The role of the team and the team members is to collect the samples needed from the locations identified in the proscribed manner and to document that they collected them in the way required. In order for any sampling and monitoring program to be implemented, a qualified team must be assembled and trained. The team should be aware of their surroundings and any observation that might affect the implementation of the SMP. Every trip to the field should include a review of expectations, logistics, and standard (and not so standard) procedures to be used for this sampling effort. Personnel conducting field sampling and monitoring need to have the appropriate qualifications for the field work as listed in the sampling plan, including physical capability, training, formal education, experience, and a satisfactory knowledge of the requirements of the activities to be carried out. Of special importance is health and safety training and that communication and documentation systems are operable, robust, and reliable.

Training should be sufficient to ensure that the staff can perform their duties competently and safely. The most important elements of training are about including the team members (those who are collecting the data in the field or in the laboratory) in the monitoring effort in a way that shares the understanding of the objectives, promotes personal responsibility for the reliability and usefulness of the data they generate, and creates a sense of ownership and participation of/in an important process. The whole team should review the SMP (each member should have his or her own copy), protocols, a health and safety plan (specifics should be reviewed the morning of sampling), and the roles and responsibilities of each team member.

Systematic training and SOPs improve the reliability and comparability of the data collected. For example, a technical protocol can be written for instructing the user how to perform a specific analytical method to be followed in the laboratory or field (such as field testing using a pH meter), how to collect a sample in order to preserve the sample integrity and representativeness (such as collection of samples for future analysis of volatile organic compounds or trace metals), or how to conduct a bioassessment of a freshwater site.

Another part of the team is the laboratory. It is good practice to maintain effective communication with the laboratory and analysts. Additional information about samples that might be helpful to the laboratory staff includes sample type, sample description, preservation method(s), a complete list of analytes and requested analytical methods, a list of possible interferences, the date delivered to the laboratory, and contact information for a knowledgeable person who can provide additional information about the samples and sample-collection methods. Selected sample collection methods are briefly described in Appendix 3.

6.2.3 Define Field Operating Procedures—Protocols/Standard Operating Procedures

Protocols or SOPS are written procedures or instructions of practices and methods used in sampling, monitoring, analytical techniques, and data management (USEPA 2007). Standard protocols and SOPs are important because they provide scientists, managers, stakeholders, and other parties involved with the project, including newcomers, with technical information on how particular jobs are performed. The SOPs can be written by a variety of entities, but ultimately they must be incorporated into the sampling plan and approved by those signing off on the plan. This step is very important because standard protocols may not be appropriate or adequate in all situations. Therefore, protocols should be available to the team prior to sampling and also available in the field for consultation if questions arise. Protocols include methods of sampling, sample collection, sample preservation, interference mitigation, sample preparation, sample decomposition, sample analysis, sample storage/archiving/disposal, data analysis and interpretation, and data management/archiving. Standard protocols improve communication between scientists and other parties and achieve common goals. Discussion of some of these methods can be found in Appendix 3. The term *standard operating procedure* is not always appropriate and the term *protocol* can generally be used. Protocols are typically media specific and often are formalized as SOPs. It is always important to document any deviation from these SOPs. Topics that need to be covered in the SOPs would be sampling procedures, including sample containers, volume, interference mitigation, preservation, cleaning and decontamination of equipment and containers, field equipment calibration, maintenance, testing, inspection, and field documentation. Procedures should also be covered on how to perform the task, wherein each step is detailed. Flowcharts are especially helpful. USEPA has a guidance document for preparing SOPs (USEPA 2007) and the Dairy Farm Business (2001) has an excellent guide to writing SOPs.

6.2.3.1 Field Data Collection Information

Specific information needs to be documented for each sample when it is collected in the field. This information includes identification number, county and state, date and time of sampling, name of person collecting sample, sample media, sample type (e.g., composite, grab), sample description, sampling method, sampling equipment, sampling location (e.g., Global Positioning System [GPS], Universal Transverse Mercator or mine coordinates, latitude/longitude from a map and a photo), sample location on study site, depth or elevation at collection site, purpose of collecting sample, observational field information (e.g., temperature, weather conditions, site-specific information, collection problems), field measurements, constituents to be determined, instrument calibration information, and preservation methods (USEPA 2008). Additional site description information could include mining district, mine name, commodities recovered, geochemical overview (main geochemical issues at the site including known data, previous spill of process tailings facility, and discharge of seeps), description of catchment basin and size, landscape/topography, land use, predominant bedrock lithology, vegetation, photos of landscape and site, and a hand-drawn map of the site. It can be quite useful to compile meteorological information (e.g., temperature minimum/maximum, precipitation, and snowfall) at the time of sample collection and to retain that information with the analytical information. Photographs are always useful to document sampling sites and conditions for future consideration. Appendix 5 includes descriptions of field data collected for specific sampling and monitoring case studies.

6.2.3.2 Field Equipment

Field equipment should be adequate for the tasks, and maintained and calibrated to provide accurate measurements. To keep track of each piece of equipment, a unique identification number can be linked to manufacturer maintenance and calibration protocols, and operational instructions. The measurement equipment should be calibrated before being put into service and thereafter according to an established procedure, including maintaining records of all equipment calibrations and maintenance (USEPA 2013).

6.3 SAMPLING DESIGN AND THE SAMPLING PLAN DOCUMENT

The sampling design should always be focused on what data are needed to answer questions and associated data quality objectives. Chapter 3 expands on the special considerations for development of the sampling plan design for mining and processing sites. A useful reference is USEPA Method QA/G-5S, *Guidance on Choosing a Sampling Design for Environmental Data Collection for Use in Developing a Quality Assurance Project Plan* (USEPA 2002c). Sampling locations and the rationales for the selection should be targeted and recorded on a map and table within the sampling plan. A study area reconnaissance is always advised. The spatial boundaries identified for the investigation should also be delineated on the map. Timing of sampling at each location should be considered and specifically programmed if it could influence the objectives and the reliability of the data.

For example, in one SMP for a mining site, sample locations were selected based on the results of previous sampling efforts at this site and existing data needs in order to maintain data comparability. The purpose was to assess current surface water and sediment conditions and to further determine the nature and extent of the environmental impacts. The data collected were to be used to evaluate the toxicological impacts and provide an ecological risk assessment.

Site locations are often selected to generate a background or reference data set, to determine source contribution to existing levels, and to evaluate potential contributions from other sources (i.e., tributaries). Additional sampling sites may be listed in the sampling plan to be added in the field based on observed conditions. (See Sampling and Analysis of Mountain Waters sidebar.)

To address sampling parameters, the following data might be collected for a typical surface water sampling event for a mine in operation:

- Field water-quality parameters—pH, conductivity, dissolved oxygen, temperature, GPS locations, and photos (as needed)
- Volumetric stream flows—using flowmeters, flumes (where necessary), and observation of stream flow measurement from fixed instrumentation
- Representative surface water samples—including streams, seeps, and all discharges for dissolved metals and total metals
- Sediment—total metals and total mercury
- Macroinvertebrates—assemblage (species identification and count)

For example, based on the conceptual model of the previously mentioned hypothetical operating mine and past sampling verification in the same area, the contaminants and sources that potentially pose a risk to ecological receptors, based on site-specific water quality standards and the watershed pathways, require further characterization. The primary analytes of concern identified for the site are aluminum, cadmium, calcium, copper, lead, manganese, nickel, silver, and zinc. However, the full suite of metals will be analyzed because additional potential contaminants may exist and the additional incremental cost for the complete suite is negligible. The process for modeling ecological risks was acknowledged and the appropriate parameters for each ecological receptor identified. No statistical or quantitative decision rule could be established prior to sampling. When the data are analyzed, if the calculated hazard quotient does not exceed one for acute toxicity, there is no ecologically significant effect observable in direct studies of community structure/function compared to reference, and site-specific toxicity tests do not show significant mortality effects compared to reference and laboratory control, then remedial actions are not likely to be necessary. Tolerable limits on decision errors, which are used to establish performance

SAMPLING AND ANALYSIS OF MOUNTAIN WATERS—CHALK CREEK

Carol C. Russell, U.S. Environmental Protection Agency

Chalk Creek in central Colorado drains an area of approximately 250 km^2 (24,864 ha), including the primary source of acid rock drainage (ARD), the Mary Murphy mine. Lode mining developed on the valley slopes producing from gold, silver, and lead-zinc deposits through an extensive system of underground workings on numerous levels along sulfide veins. The Golf Tunnel was constructed to transport ore to the mill and to provide drainage to the 670-m (2,198-ft) (vertical) mine workings of the Mary Murphy mine. Work by the Colorado Division of Minerals and Geology (CDMG), their consultant, and the U.S. Environmental Protection Agency involved the characterization of the ARD groundwater flow systems. Multiple dyes were introduced into the fractured bedrock mountain/mine pool system. Results from the tracing studies indicate four discrete groundwater flow systems: (1) a relatively slow-flow system in the high mountain faulted bedrock very high in zinc; (2) rapid flow through the cavern-type systems of the interconnected mine voids and fractures lower in the mountain; (3) a weathered, shallow fracture system in the upper 8 m (26 ft) of bedrock; and (4) infiltration of precipitation through the tailings repository into the near surface system, also very high in zinc (INSTAAR n.d.). After examining one of these adits, mining specialists from CDMG found that ARD was draining from only a few fractures in the rock. In fact, most of the mine drainage originated from the very back of the Golf Tunnel from a shaft drilled vertically to intercept the mine workings some 245 m (804 ft) above. Very little ARD was draining into the tunnel except from this site, even though many other sulfide vein systems were mined. Mine inflows were segregated into clean water that could be piped out directly without treatment and the one ARD site, which can be readily collected for passive treatment. The major lesson learned is that if uncontaminated water can be intercepted and diverted before it enters mineralized areas of mine workings and becomes ARD, the total loadings of metals will be significantly diminished (Russell 2006).

goals for the data-collection design, need to be established. For this example, the number of samples and station locations are based largely on past investigations. Some sampling locations had been eliminated based on evaluation of their contribution to the data set as a whole.

6.4 CONDUCT THE PILOT STUDY

Conducting a pilot study, pre-study, screening test, feasibility study, or orientation survey consists of a series of preliminary experiments prior to conducting the final plan to work out any difficulties or problems that could arise, in order to improve quality and efficiency. A pilot study or the first step of a phased monitoring program can reveal deficiencies or problems with the procedures or design of a plan before time and resources are expended on large-scale plans. Using a phased approach is suitable for sampling to get a preliminary understanding of the problem, variability in the data and types of materials (waters, sediments rocks), and then allows subsequent phases to build on early phases with the goal of reducing uncertainty filling data gaps and conducting more sampling of those areas with the greatest uncertainty and variability. A pilot study can also address logistical issues, such as evaluating if technicians have the proper skills and training, if the equipment operates correctly under field conditions, if QA/QC procedures are adequate, if statistical applications are sufficient, and identifying any other difficulties under field conditions. A pilot study allows for testing of the hypotheses based on the conceptual model and can lead to testing additional hypotheses. After conducting the pilot survey, it is important to consider what worked and modify the plan accordingly, including the protocols, if necessary.

For another sampling effort, the following text describes an example of an operating mineral processing facility. The conceptual model and past monitoring identified plumes of impacted groundwater. The model is being used to analyze and compile new data and generate three-dimensional visualizations of the aquifers and associated plumes. The conceptual model and data collection will be continuously updated until the plume has been fully delineated horizontally and vertically and all data gaps have been filled. The sampling plan consists of borehole screening sampling of groundwater. If possible, concentrations of target analytes (e.g., arsenic and trichloroethylene) in groundwater during screening-level sampling will be determined by an on-site mobile laboratory. Project screening levels estimated from the preliminary human health risk assessment would be used to delineate the plume vertically and horizontally. The screening samples would also be compared to regional action levels. The results of the screening-level analysis will be used to determine offset boring locations and well installation locations. The borehole screening samples to be collected and analyzed would be used for field decisions and well location selection, and these screening results will possibly be used as additional information for the final risk assessment.

6.5 IMPLEMENT THE SAMPLING AND MONITORING PLAN (PROGRAM IMPLEMENTATION)

Three basic steps are involved in implementing the SMP (program implementation): sampling, analysis, and interpretation. These appear to be independent, but in reality, they are interdependent functions. If problems are encountered at the sampling step, all efforts in the succeeding steps may fail. The sampling stage is probably the most important and most difficult to correct once the samples are collected. Therefore, it is imperative that the initial stages previously discussed are well thought out and implemented before samples are collected. Field personnel must adhere to all appropriate protocols.

Implementation of health and safety procedures while sampling is of utmost importance. Knowing the hazards and actions to take if something goes wrong should be identified in the planning stage and carried out in the implementation phase. For instance, sampling with a partner rather than solo sampling has saved numerous lives. Also, knowing the location of the nearest hospital or emergency services can save time as well as lives.

6.6 SUMMARY

Implementation of the SMP consists of tasks for management and leadership of the sampling effort, assembly and training of the monitoring team, identification of field data collection information, including the sampling design, protocols and SOPs, and conducting the pilot and full-scale study. Additional items that could be included in an SMP include a formal statement of the purpose of the sampling, a formal statement of assumptions, a list of the parties responsible for implementing each step of the plan, a signed confirmation of participants and their responsibilities in the plan, a formal health and safety plan, a schedule of activities, a description of the QA/QC procedures, and documentation of arrangements made for access to sampling locations.

6.7 REFERENCES

Dairy Farm Business. 2001. *Standard Operating Procedures: A Writing Guide*. Penn State, College of Agricultural Sciences, Agricultural Research and Cooperative Extension. http://centaur.vri.cz/news/prilohy/pril771.pdf. Accessed 9/11/13.

INSTAAR (Institute of Arctic and Alpine Research). n.d. *Mary Murphy Mine—Characterizing Acid Mine Drainage Flowpaths and Residence Times Using Isotope Tracers.* http://snobear.colorado.edu/Daniel/mineresearch/mmm.html. Accessed 9/11/13.

Russell, C. 2006. Mine drainage challenges from A to Z in the United States. In *Proceedings, 7th International Conference on Acid Rock Drainage (ICARD 7).* Lexington, KY: American Society of Mining and Reclamation. pp. 1785–1802. www.imwa.info/docs/imwa_2006/1785-Russell-CO.pdf. Accessed 9/11/13.

USEPA (U.S. Environmental Protection Agency). 2001. *EPA Requirements for Quality Assurance Project Plans (EPA QA/R-5).* U.S. Environmental Protection Agency, EPA/240/B-01/003. www.epa.gov/quality/qs-docs/r5-final.pdf. Accessed 9/11/13.

USEPA (U.S. Environmental Protection Agency). 2002a. *Guidance for Developing Quality Systems for Environmental Programs (EPA QA/G-1).* U.S. Environmental Protection Agency. EPA/240/R-02/008. www.epa.gov/quality/qs-docs/g1-final.pdf. Accessed 9/11/13.

USEPA (U.S. Environmental Protection Agency). 2002b. *Overview of the EPA Quality System for Environmental Data and Technology.* U.S. Environmental Protection Agency. EPA/240/R-02/003. www.epa.gov/quality/qs-docs/overview-final.pdf. Accessed 9/11/13.

USEPA (U.S. Environmental Protection Agency). 2002c. *Guidance on Choosing a Sampling Design for Environmental Data Collection for Use in Developing a Quality Assurance Project Plan (EPA QA/G-5S).* U.S. Environmental Protection Agency. EPA/240/R-02/005. www.epa.gov/quality/qs-docs/g5s-final.pdf. Accessed 9/11/13.

USEPA (U.S. Environmental Protection Agency). 2006. *Guidance on Systematic Planning Using the Data Quality Objectives Process (EPA QA/G-4).* U.S. Environmental Protection Agency. EPA/240/B-06/001. www.epa.gov/quality/qs-docs/g4-final.pdf. Accessed 9/11/13.

USEPA (U.S. Environmental Protection Agency). 2007. *Guidance for Preparing Standard Operating Procedures (SOPs) (EPA QA/G-6).* U.S. Environmental Protection Agency. EPA/600/B-07/001. www.epa.gov/quality/qs-docs/g6-final.pdf. Accessed 9/11/13.

USEPA (U.S. Environmental Protection Agency). 2008. *Field Measurement Uncertainty.* U.S. Environmental Protection Agency. SESDPROC-014-R0. www.epa.gov/region4/sesd/fbqstp/Field-Measurement-Uncertainty.pdf. Accessed 9/11/13.

USEPA (U.S. Environmental Protection Agency). 2013. *Field Operations Group, Operational Guidelines for Field Activities.* EPA report. www.epa.gov/region8/qa/FieldOperationsGroupOperationalGuidelinesForFieldActivities.pdf. Accessed 9/11/13.

CHAPTER 7

Data Management, Assessment, and Analysis for Decision Making

Carol C. Russell, Kathleen S. Smith, and Virginia T. McLemore

7.1 INTRODUCTION

The aim of a sampling and monitoring program is to inform the decision makers about minimizing and mitigating the potential harm, and to enhance the potential benefits of the project based on scientific data. Good quality data managed properly serve as the core to all subsequent design, technical analysis, and responses and need to be conducted under proper quality assurance procedures, well organized, and available for analysis. This chapter describes steps to ensure that data are managed appropriately and provides basic data analysis strategies. Specifically, monitoring helps to determine and implement timely remedial measures to ensure that environmental impacts are prevented or minimized. In addition, monitoring can measure if actions taken have achieved the desired results and that resources have been utilized appropriately, efficiently, and effectively. For monitoring to achieve these expectations, the data must be analyzed, assessed against a goal, managed, and presented in such as way as to inform the decision-making process. Decisions are made not just on the information provided but on other factors regarding values, relationships, and the integrity of the process.

7.2 DATA MANAGEMENT

Data and information management is a process that starts with project design and can extend well beyond the data analysis and information dissemination phases. Management, interpretation, and presentation are highlighted to convey that data have little intrinsic information and must undergo processing to be transformed into information. Data are like letters of the alphabet: taken individually, they reveal very little; put together with a little thought and organization, however, those same letters can tell a complete story.

The amount of data and information collected at mine sites is extensive, thus requiring the need for effective data management to enable the mining company as well as federal, state, and local agencies and other stakeholders to make informed decisions. To make the best use of baseline data collected during the exploration phase, the data need to be available to mine planners and metallurgists, and also need to be communicated to the wastewater treatment managers. Requirements for data management and sharing need to be identified before sampling and monitoring begins.

Data handling and the use of databases is a major undertaking throughout a monitoring program. Six basic components of data management are

1. Project design and planning,
2. Data management (paper and digital),
3. Capture of metadata,
4. Data archiving,
5. Data manipulation (statistics), and
6. Presentation of information.

Efforts are being made to develop data storage systems that facilitate data exchange among different groups that conduct sampling and monitoring.

A *database* is a system where data are stored, whereas a *relational database* is a database where the data are organized into similar entities that are related to one another by a common, unique identification number (called a primary key). Relational databases facilitate effective management of a large amount of investigation data by eliminating duplication and maintaining consistency of the data. Databases can be developed for a specific resource (e.g., surface water) or as a general clearinghouse of information. One example of a broad database is the U.S. Environmental Protection Agency's (USEPA's) national water and biological data storage and retrieval system, called STORET. A specific set of quality control measures (metadata) is required for any data entered into the system. For more information, see Section 7.8 (USEPA 2011). Databases are best if they are designed or modeled before or during the first stage of the sampling and monitoring program. The typical mining-related data to be incorporated into a database or a set of relational databases include locational data, mineralogy, chemical analyses, geologic data on drillholes and test pits, historical and recent photographs, and other data. It may be useful to coordinate mining and environmental data into one relational database. One of the most effective tools to enter, store, report, and utilize these data are by relational databases such as SQL Server, Access, FoxPro (all by Microsoft, Redmond, Washington); Paradox (Corel, Ottawa, Ontario); GIS (geographic information system) databases; or on other servers (McLemore et al. 2004). Relationships among data are stored in the database along with the data and the definitions of the data, called metadata.

Data can be stored, processed, and retrieved using four major components to transfer and exchange data: (1) database management systems, (2) word processing software, (3) computer-aided drafting or CAD software, and (4) GIS software. Computer technology is such that all of these techniques can be connected or linked together. Within a database management system, the tables can be linked to each other, where appropriate, and a main table can provide the organizational structure to other linked tables. The database is linked to additional data in other file formats outside of the database, such as photographs, reports, laboratory scans, and so on. Database management systems query and manipulate tabular data for further calculation, analysis, reporting, or exchange. Some databases can be designed to display statistical data and graphs automatically.

Geographic or locational data includes both surveyed points and GPS (Global Positioning System) data. CAD software applies computer graphics to manual drafting techniques and includes design, drafting, and display of graphically oriented data. GIS software captures, manages, manipulates, analyzes, models, and displays spatial (locational) data. Once the data are entered into appropriate databases with locations, the data can easily be converted into GIS format for displaying on maps and performing geospatial analyses.

The fields used in the database must be fully explained and documented, which is referred to as metadata. Metadata documents need to contain documentation on methods used, sources of data, data quality, and any pertinent information that documents the data collected, such as special handling procedures (see Metadata sidebar). It is important that staff is assigned to input and maintain the database throughout the project.

Everything needs to be documented, from field studies to modeling runs. Computers are ideal to electronically capture, store, distribute, annotate, display, and disseminate the extensive data collected at mine sites. Some mining and consulting companies have developed sophisticated computer programs to handle these data. Other companies use commercially available software. Capability with multiple interrelated systems needs to be considered. Evaluating data by simply inputting it into specially designed software is not a recommended method because the oversight of a competent person is invaluable. Computers are merely tools to aid in evaluation and interpretation of the data.

The most difficult task is maintaining quality control (QC) of the data being entered into the database and the integrity of the database itself. QC for data management is a system of procedures, checks, audits, and corrective actions to ensure that data are collected, stored, and analyzed in an acceptable manner, and that technical aspects and reporting are adequate.

7.3 ANALYSIS AND ASSESSMENT OF DATA

The term *scientific assessment* means interpretation of data based on scientific or technical knowledge from multiple factual inputs, data, models, assumptions, and/or professional judgment to bridge uncertainties in the available information. Different types of assessments are performed based on the objectives of the sampling and monitoring effort. The four general types of assessments include

1. Condition assessments to detect chemical, physical, and biological status;
2. Causal pathway assessments to determine causes and identify their sources;
3. Predictive assessments to estimate environmental, economic, and societal risks and benefits associated with a range of possible management actions (Cormier and Suter 2008); and
4. Outcome assessments to evaluate the results of the decisions made.

Condition assessments ask, Is there a problem? Causal pathway assessments ask, What caused the problem? Predictive assessments ask, What will be the consequences of taking a particular action? Outcome assessments ask, Did the solution work?

In a similar fashion, an environmental assessment undertaken for a mining operation would help to support decision making to protect the environment. The assessment puts a value on something to determine its positive or negative value. At a mining site, assessments of impacts are either predictive, as is the case of environmental impact assessments or after-the-fact assessment of damage, normally called a site assessment. A good reference for abandoned mine assessment is USEPA's *Abandoned Mine Site Characterization and Cleanup Handbook* (USEPA 2000a).

The assessment is broken down into human or environmental risk assessments, as discussed in Chapter 4. Another gauge of impact might be the exceedance of a standard such as a drinking water standard established in the Safe Drinking Water Act or the exceedance of a screening level for soil cleanup utilized under Superfund.

7.3.1 Evaluate Laboratory Results

Quality decisions are based on quality information derived from quality samples (data). However, it is often difficult to obtain quality and reproducible samples. The best "tool" in the production of quality samples is to systematically avoid errors at the beginning rather than trying to identify

METADATA

The following data qualifiers are the minimum information about a data point that is necessary to ensure consideration for standards and classifications proceedings.

A. **Project description:** Reasons why data was collected.

B. **Data source:** Who supplied the information or from what database was information taken?

C. **Waterbody type:** Type of waterbody sampled (stream, lake, reservoir, wetland, ditch, drain, etc.).

D. **Location:** A geographical location must be supplied with the data. The location description must contain enough information for the sample point to be located on a 1:24,000 quadrangle map. The sample location would best be described by latitude and longitude to at least the nearest second. When using a geopositioning device (e.g., GPS), the geopositioning method and horizontal datum must be noted. Other location descriptors include state, county, and hydrologic unit code.

E. **Date and time:** The date and time of sample collection should be noted. On time-composite samples, the beginning and ending times should be supplied.

F. **Characteristic measured:** This would include the name of the characteristic and its sample fraction (e.g., ammonia as nitrogen, dissolved iron, suspended solids).

G. **Result value measured:** This includes the quantitative or numeric result measured. This could also be a qualitative result such as good, fair, or poor. The quantitative value must have reporting units in metric (International System of Units).

H. **Quantification and/or detection limit:** This is the quantification limit (QL) and/or method detection limit (MDL) as reported by the laboratory for each analysis. The MDL is the smallest amount of a substance that can be detected, although not necessarily quantified, by a particular method. The QL is the smallest amount of a given substance that can be detected and quantified at a level that is statistically significant. The QL is sometimes referred to as the practical quantification limit, or PQL. The PQL is approximately 3 to 10 times the MDL but is typically defined as a quantity that is 5 times the MDL.

I. **Method techniques:** This is information about whether the sample was grab or composite. If composite, additional descriptors such as time, flow, width, depth, or depth weighted should be noted. Instrumentation is used to collect and analyze the sample. Information on handling and transportation of the sample is also required.

J. **Who measured:** Information should be supplied as to who physically collected the sample and who did laboratory and field analyses. This can be by agency, company, laboratory, or individual.

K. **Lake/Reservoir information:** Most lake and reservoir monitoring plans include collecting depth profiles for temperature, dissolved oxygen, pH, and specific conductance. Complete profile information should be included. Results for each meter depth should be provided for each sampling date. Depth information for water quality samples should include upper and lower depths for composite or integrated water column samples, or as the actual depth for discrete depth grab samples.

L. **Other:** Although not required, it is suggested that quality descriptors such as method precision and bias be included with value or data set. Also, inclusion of reference and method number for the analytical method would add greater specificity as to the chemical form being measured.

Source: Adapted from USEPA 2013

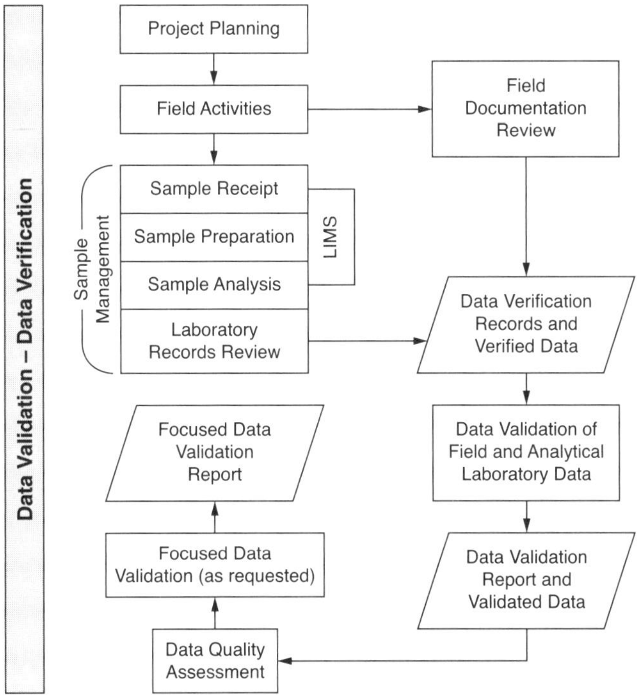

LIMS = Laboratory Information Management Systems.

Source: USEPA 2002.

FIGURE 7.1 Data verification and data validation components in the project life cycle

problems with the data after the fact. Figure 7.1 illustrates steps in the verification and validation of data.

Nevertheless, laboratory results can be evaluated in a number of ways to determine if the data are useable (Weight and Sonderegger 2001). Tools can be used to minimize uncertainties and maximize confidence in decision making. Comparison between field and laboratory measurements, ionic balances, spikes, sample splits, and additional laboratory results are common indications of precision and also improve QC. However, these approaches do not guarantee accuracy, and, ultimately, the analyses must make geologic sense and compare with the known mineralogy and geologic setting. Different laboratories may use the same analytical methods on the samples, but the results could still be significantly different. Laboratory results should be compared with field data collected, which can identify potential problems of sample stability, preservation methods, poor equipment, or poor technique. Field results may be temperature dependent, so it is important to record the temperature of the sample in the field, especially for water samples. In the laboratory, samples must be analyzed at room temperature, as volumetric glassware is calibrated at room temperature.

Determining ionic balances is another method of evaluating laboratory results for water quality samples. Typically, analyses of water samples include major anions and cations. The sum of the milliequivalents per liter of the major anions should equal the sum of the milliequivalents per liter of the major cations. Solid samples should include major components: SiO_2 (silicon dioxide), TiO_2 (titanium dioxide), Al_2O_3 (aluminum oxide), CaO (calcium oxide), MgO (magnesium oxide), MnO (manganese (II) oxide), K_2O (potassium oxide), Na_2O (sodium oxide), P_2O_5

(phosphorus pentoxide), and loss on ignition or LOI. The sum of the major components should total 100%. The sum of anions will not always equal the cations and the sum of major oxides will not always equal 100%, and the acceptable deviations must be agreed upon.

Samples can be spiked in the field or in the laboratory to check the accuracy of analytical measurements. A *spike* is a controlled addition to a sample of any media with a known amount of the analyte(s) to be determined.

Sample replicates and splits are recognized and highly recommended methods of determining the variability of the results. In the field, sample replicates include duplicate samples collected at the same site but given different sample numbers. Large samples collected at the site can be homogenized and split to obtain additional samples to test for precision and variability. During the sample preparation, sample splits are critical in determining if the preparation methods are appropriate. During the analysis, sample splits are regularly performed to check procedures and proper functioning of the instrument.

It is good practice to send splits of the same samples to different laboratories for comparison and to include certified samples (with similar sample numbers as the field samples) with the spikes. Major elements should agree within ±5%, and trace elements should agree within ±10% (Weight and Sonderegger 2001).

7.4 DATA PRESENTATION

The goal of data presentation is to identify patterns and trends in the data that could go unnoticed using purely numerical methods. In addition, visual tools can be used to effectively convey information to nonscientist decision makers and stakeholders. People tend to understand pictures much more readily than mathematical formulas. Graphs can be used to identify patterns and trends, to quickly confirm or disprove hypotheses, to discover new phenomena, to identify potential problems, and to suggest corrective measures. Graphical representations include displays of individual data points, statistical quantities, temporal data, and spatial data. Because no single graphical representation provides a complete picture of a data set, the analyst should choose different graphical techniques to illuminate various characteristics of the data. At a minimum, the analyst should choose a graphical representation of the individual data points and a graphical representation of basic statistical properties. An example of discontinuous data is the number of samples collected on a specific date. In contrast, an example of continuous data would be measurement of flow as a function of time.

Numerous options for data presentation are described as follows:

- A bar chart has a vertical and horizontal portrayal of information with the rectangular bars being proportional in length to the values they represent. Bar charts are used for plotting discrete (or "discontinuous") data (i.e., data that have discrete values and are not continuous). Comparing the monitoring results to the goals established in the sampling and monitoring plan is very useful. An example is a comparison of concentrations of metals in water with the appropriate water quality criteria.
- A histogram is a graphical representation that looks similar to a bar chart but consists of tabular frequencies or probability distribution of a continuous variable (Figure 7.2). The categories are usually specified as consecutive, non-overlapping intervals of a variable. The categories (intervals) must be adjacent and are often chosen to be of the same size. For example, the number of samples by month that have a pH of less than 5 can be portrayed in a histogram.

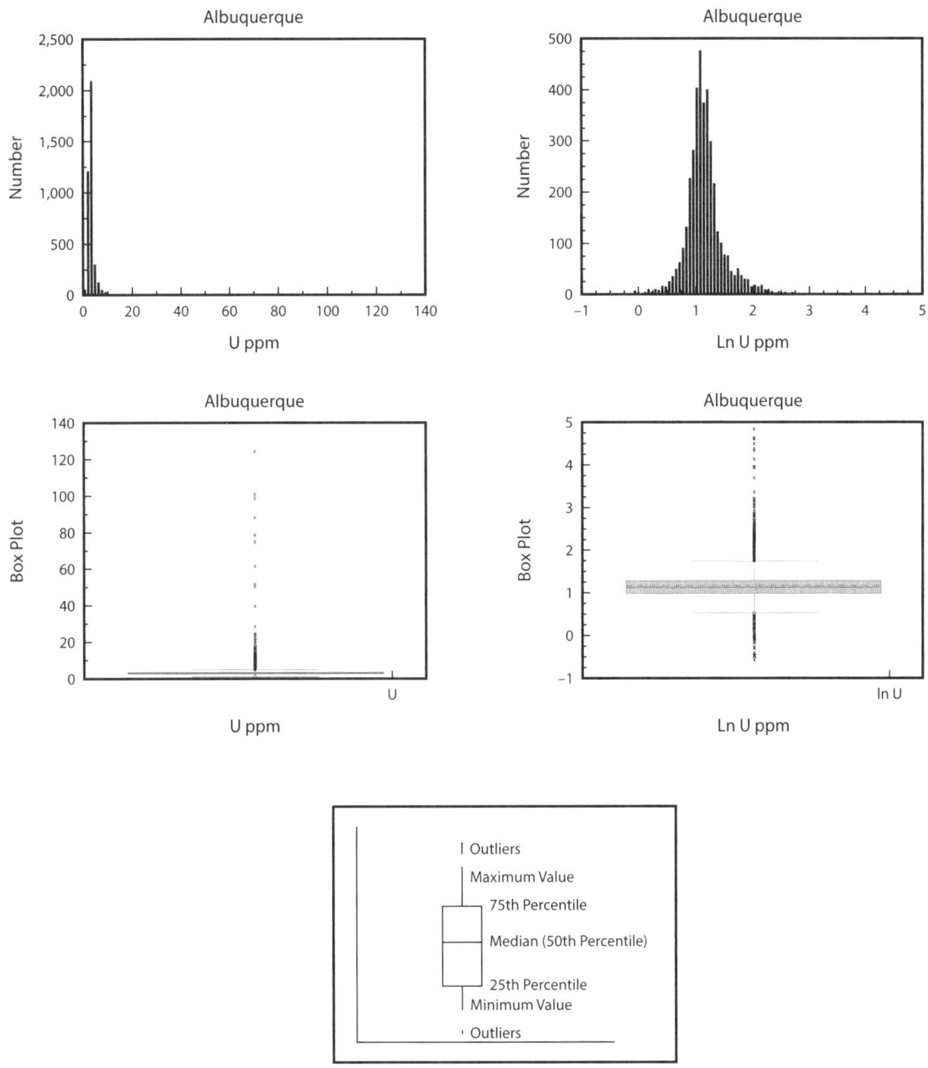

Source: McLemore 2010.

FIGURE 7.2 Histograms and box plots of uranium analyses in stream sediments for the Albuquerque 1:250,000 topographic quadrangle

- A box plot (also known as box-and-whiskers plot) is composed of a central box divided by a line with two lines extending out from the box called whiskers (Figure 7.2). The length of the central box indicates the spread of the bulk of the data (the central 50%), whereas the length of the whiskers show how stretched the tails of the distribution are. The width of the box has no particular meaning, so the plot can be made quite narrow without affecting its visual impact. The sample median is displayed as a line within the box, and the sample mean is displayed using a plus (+) sign. Any unusually small or large data points are displayed by an asterisk (*) on the plot. A box plot can be used to assess the symmetry of the data. If the distribution is symmetrical, then the box is divided in two equal halves by the median; the whiskers will be the same length and the number of extreme data points will be distributed equally on either end of the plot.

7.5 DATA INTERPRETATION

Although computers are quite helpful in organizing data, deciphering the story behind these facts remains a human task. The overall purpose of data interpretation is to get answers to study design questions—the same questions that originally provoked the start of a monitoring program. The products of risk assessment (interpretation) can be thought of as, among other things, communication products. Their value lies in their contribution to the objectives of the decision-making function, including their effects on the primary decision maker and other interested parties who participate in the decision or otherwise use the information that the products convey. As in any complex design problem, the process of design is intended to find the best solution to achieve multiple simultaneous and competing objectives while satisfying constraints on the process or the end product. One will need to avoid *confirmation bias*. As the term is typically used in the psychological literature, it connotes the seeking or interpreting of evidence in ways that are partial to existing beliefs, expectations, or a hypothesis.

Data are the raw numbers collected from the appropriate sources at the proper times using the correct methods to answer a question. This is true for all data (e.g., geologic, mineralogical, meteorological, financial, chemical, biological). These data are analyzed numerically, spatially, visually, and/or statistically and converted into information. The information received is turned into knowledge based on experience, training, and intuition; then this knowledge is used to make decisions. Without quality data, informed decisions cannot be made. Monitoring data are assessed against benchmarks, data from reference sites, or conditions, as well as by comparison with accepted guidelines for water, sediment, human health, and food quality. The development of regulatory guidelines and standards relies on published data from laboratory monitoring of human and environmental toxicology.

Interpretation of data is the process of analyzing the data in context with other information about the samples, the site, and their environment, and to understand the physical, chemical, and biological processes at work. Too often overlooked is the process of comparing the data with criteria or a standard. The planning sequence of moving from data analysis to information and knowledge is supposed to provide confidence that the sometimes costly actions to address an environmental problem are justified. The scientific method starts with limited data and information from which a tentatively held hypothesis about cause and effect is formed. The hypothesis is tested, and new understandings and new hypotheses can be stated and tested. By definition, science is the process of iterative questioning. This sampling and monitoring handbook recommends policy decisions be based on *the science*. Decisions to pursue some actions must be made, based on a preponderance of evidence, but there may be a need to continue to apply science as a process (data collection and tools of analysis) in order to minimize the likelihood of future errors.

The terms *data* and *information* are often used as synonyms, but data are not the same as information. Unanalyzed data do not constitute information. Data must be interpreted for their meaning through the filter of analytical techniques, and the result of such data analysis is information that can support decision making. Knowing what data are needed and turning those data into information constitutes, in large part, the science behind an environmental management program. Information for decision making should be based on carefully collected and interpreted monitoring data.

The techniques for transforming data into information include statistical inference methods, simulation modeling of complex systems, and, at times, simply the application of the best professional judgment of the analyst. In all these processes, there will always be some uncertainty (and thus some *unreliability*) about whether the resulting information accurately characterizes the environmental problem and the effectiveness of the solutions. Because uncertainty cannot be eliminated, determining whether the information generated from data analysis is reliable is a value judgment. Individuals and groups will have different opinions about whether and how to proceed with environmental management given a certain level of uncertainty. An optimal project design balances the uncertainties and associated costs with the resources available from the project.

7.5.1 Statistical Analysis

Statistical analysis is an important tool in summarizing, describing, analyzing, interpreting, and presenting data, and can be used as a predictive tool, but it must be used with caution and complete understanding of the data limitations and the problem to be addressed. Numerous statistical packages are available to summarize, analyze, and interpret the data (USEPA 2006, 2008). Statistical analysis of data is useful for four reasons:

1. Statistics can be used in evaluating the precision and accuracy of the data.
2. Statistics can aid in interpretation of the data and coming to justifiable conclusions.
3. Statistics can help in developing a better approach to the project, before implementing the sampling and monitoring program.
4. Statistics can be used to scientifically justify the conclusions.

It is not practical in this section to describe a rigorous evaluation and description of the advantages and disadvantages of statistical techniques. For most data, relatively simple statistical techniques will result in the most obvious interpretations. Statistics measure central tendencies and dispersion of unbiased samples. It is important to remember that a sample never defines a population perfectly; error always exists.

Once data are in hand, a number of statistical tests can be performed. Testing includes (1) tests of differences between groups (independent samples), used when a comparison is made between samples; (2) tests of differences between variables (dependent samples), used when a comparison is made between variables measured in the same sample; and (3) tests of relationships between variables (e.g., correlation coefficients). Statistical estimates can also be performed.

The choices among space, time, and thematic parameters that are made in designing a monitoring program also affect the choice of statistical analyses. Geostatistical techniques may be desired for analyzing data collected at a large number of sites. The goal of the first step in statistical analysis is to summarize some basic quantitative characteristics of the data set using common statistical quantities. Some statistical quantities useful to the analyst include

- Number of observations;
- Measures of central tendency, such as a mean, median, or mode;
- Measures of dispersion, such as range, variance, standard deviation, coefficient of variation, or interquartile range;
- Measures of relative standing, such as percentiles;
- Measures of distribution symmetry or shape; and
- Measures of association between two or more variables, such as correlation.

Many types of statistics can be used (Isaaks and Srivastava 1989; Rollinson 1993; Helsel and Hirsch 2002). Two of the most frequently used descriptors of environmental data are the mean and standard deviation. Textbook statistics commonly assume that if a parameter is measured many times under the same conditions, then the measurement values will be randomly distributed around the average with more values clustering near the average than farther away. In this ideal situation, a graph of the frequency of each measure plotted against its magnitude should yield a bell-shaped, or normal, curve. Although both the mean and the standard deviation are quite useful in describing environmental data, the actual measures often do not fit a normal distribution. Other statistics sometimes come into play to describe the data (USEPA 2006).

Normal distributions of geochemical data should not be assumed (Rollinson 1993; Helsel and Hirsch 2002) and can be determined by histograms or other statistical methods. A normal distribution is an accepted model of the behavior of certain random events and is used to approximate other probability distributions. Classical statistical analysis requires that the data are normal or log-normal and represent one population.

Descriptive statistics use indicators such as the mean, median, mode, standard deviation, coefficient of variation, and so on. The *mean* or *arithmetic average* is the sum of the values divided by the number of values. The *median* is the middle value of a list ranged from lowest to largest. The *mode* is the value of a variable that occurs most frequently, and there can be multiple modes in a set of data that is not normally distributed. The *frequency distribution* or the *frequency of occurrence* is typically illustrated by a histogram (a bar or line graph of values where the length of the line shows the frequency of that value), cumulative frequency plot, or density function plot. A *normal distribution* or *bell curve* is a density function plot where the curve is regularly shaped; other distributions of data are likely (Schreuder et al. 2004). The *standard deviation* is a measure of how much the data deviates from the mean.

Additional statistical tools are useful in sampling and monitoring programs. Regression analysis is the process used to determine the best fit of a line or curve to a set of data consisting of dependent and independent variables; linear regression is used to describe how well the trend fits along a line. Curvilinear regression is used to describe how well the trend fits along a curve. Factor analysis is a statistical tool used to group a set of many observed variables into two or more related factors. A correlation is used to define the linear relationship between two variables. Pierson correlation is obtained by dividing the covariance of the two variables by the product of their standard deviations. A correlation coefficient of 0.8 is generally a good correlation, although in some cases, lower correlation coefficients could be acceptable or indicate significant data relationships. The term *trend* is used when the data indicate a relationship, but not a strong linear or curvilinear correlation.

Outliers are the sample points that appear to be inconsistent with the remainder of the collected data (Iglewicz and Hoaglin 1993) and are important to identify and properly interpret their source. Outliers can be determined by visual inspection of scatter plots and data tables or by values greater than 1 standard deviation. Some outliers can be identified as recording errors by checking the outlier values with the original data set, such errors must then be corrected in the final data set. Any remaining outliers should be identified, and, if appropriate, described on a case-by-case basis in the project reports. In some cases, the tests or measurements can be rerun to confirm the outlier value; appropriate notes should be recorded in the database (USEPA 2000b). See Unusual Measurements sidebar.

It also is important to realize that a correlation does not imply causation or that a correlation cannot always be used to infer a causal relationship between the variables. Correlations need to be examined by defining geologic, geochemical, or geotechnical processes that could be responsible for the correlation. It is important to try to understand variations in the data. In many cases, the variation is sample heterogeneity.

Statistical analyses can be used for description, communication, and to test hypothesis regarding the population from which the data were drawn. The confidence interval estimates the range in which a population parameter falls. A confidence interval is a range of a probability

UNUSUAL MEASUREMENTS CAN BE VALID—IRON MOUNTAIN MINE

Carol C. Russell, U.S. Environmental Protection Agency

From the 1860s through 1963, the Iron Mountain mine site in northern California was periodically mined for iron, silver, gold, copper, zinc, and pyrite. Many miles of streams, tributaries, and reservoirs within the Sacramento River Basin are impacted by the acid drainage (AD) from the mine with a pH level measured at −3.6. Salmon kills have been noted since 1899. Releases of stored AD into the reservoirs are timed to coincide with the presence of diluting releases of water from Shasta Dam. One option for remediation was to plug the mine portal. Studies predicted that if the mine portals were plugged, a pool of water would develop with extremely acidic pH (near 1) and high metal concentrations (grams of dissolved metals per liter). The flooded mine pool would have a volume of approximately 600,000 m^3 and would present a high risk to the environment and to the local community because of the potential for catastrophic plug failures, faulty plug seals releasing mine water, and acid seeps developing in other locations. Based on this information, plugging of the mine was dropped from consideration. Instead, a combination of remediation activities was chosen with the clear understanding that new technologies are needed. The options chosen were an acid neutralization treatment plant (lime/sulfide high-density sludge treatment process), capping of selected areas, and surface water diversions. An agreement regarding the payment to state and federal governments of close to $1 billion has been structured using a unique funding mechanism of combined financial assurance and insurance (USEPA 2010, 2011).

distribution and does not provide any information about individual values in the population. Hence, confidence intervals cannot be used in applications such as comparison to environmental standards, because the confidence interval estimates a population parameter (e.g., mean) but does not provide information about individual values. The tolerance interval estimates the range that should contain a certain percentage of each individual measurement in the population. Tolerance intervals tell us what the individual values should be and are especially useful in compliance monitoring when one is concerned with maximum contaminant levels.

The two significant sources of uncertainty in any water quality management program are epistemic and aleatory uncertainty (Stewart 2000). *Epistemic uncertainty*, incomplete knowledge or lack of sufficient data to estimate probabilities, is a result of reliance on models relating sources of pollution to human health and biological responses. Our conceptual understanding of the systems under study is incomplete, because models are necessarily simplified representations of the complexity of the natural systems and are derived from limited data for simulating the systems. Thus, the goal should be to increase the availability of data, improve its reliability, and advance the modeling capabilities (NRC 2001).

Complete certainty in support of environmental management decisions cannot be achieved because of *aleatory uncertainty*, the inherent variability of natural systems and processes. Aleatory uncertainty arises in systems characterized by randomness. Not only is the environment characterized by randomness, but it is also an open system in which the boundaries of possible biological outcomes cannot be known in advance.

7.5.2 Spatial Interpretation Methods

Spatial interpretation or analysis includes any technique using topological, geometric, and geographic properties. In mining and environmental science, spatial analysis generally involves different techniques that show samples and resulting data on maps. Techniques such as variography, contouring, bubble plots, and kriging are useful in determining optimal sampling patterns for the area and to interpolate values in between sample points. GIS technology has evolved into a powerful tool in evaluating and analyzing large, complex data in the mining, environmental, and geological fields. ESRI ArcMap is a part of ArcGIS and includes a suite of geospatial processing programs that allows the viewer to examine, plot, edit, and analyze geospatial data and aid in establishing relationships between complex sets of data; see Section 7.8). Integrating numerous data sets is an important tool in mine reclamation sites (Grunsky 1997; McLemore 2010). Pattern recognition is a key concept in data analysis, and a variety of techniques are available to evaluate the geographic representation of the data. The R Project for statistical computing (see Section 7.8) provides another technique for evaluating spatial data. In many instances, statistical analysis of the data is the only way to interpret the large quantity of data collected. However, special problems may be encountered when statistical techniques are applied to geological and geochemical data because of the heterogeneity of the earth's crust. Geostatisticians are constantly refining statistical techniques to better account for these heterogeneities.

Geostatistics is a branch of applied statistics concerned with spatial data and can be very useful in interpreting environmental data. In geostatistics, each data value is associated with a location in space. *Semivariance* is a measure of the degree of spatial dependence between samples. The plot of the semivariances as a function of distance from a point is referred to as a semivariogram. Use of geostatistics in data analysis impacts sampling methods. In designing a sampling plan intended for geostatistical interpretation, there is no need to avoid autocorrelations; hence, sampling becomes less restrictive. Use of geostatistics changes the emphasis from estimation of averages to mapping spatially distributed populations. *Kriging* is a geostatistical technique to interpolate the value at an unobserved location from observations of its value at nearby locations.

7.5.3 Models

Models can be very useful tools for data interpretation. For example, the conceptual model described in Chapter 5 can be used to test a hypothesis that is later confirmed, or not, through data collection. A predictive model is broadly defined to include both mathematical expressions and expert scientific judgment. Mathematical models can be characterized as *empirical* (also known as statistical) or *mechanistic* (process-oriented), but most useful models have elements of both types. An empirical model is based on a statistical fit to data as a way to statistically identify relationships between stressor and response variables. A mechanistic model is a mathematical characterization of the scientific understanding of the critical biogeochemical processes in the natural system; the only data input is in the selection of model parameters and initial and boundary conditions.

Data availability is one source of uncertainty in the development of models for decision support. Although there are other sources of uncertainty as well, models should be selected (simple versus complex) in part based on the data available to support their use. The use of complex mechanistic models is warranted if it helps promote the understanding of complex systems and as long as uncertainties in the results are reported and incorporated into decision making. However, there may be a tendency to use complex mechanistic models to conduct water quality assessments in situations with little useful water quality data and/or involving major remediation expenditures or legal actions.

7.6 DECISION-MAKING PROCESS

In general, decision making follows a six-step format:

1. Formulate the problem (question) to be answered,

2. Analyze the risks,
3. Define the options or alternatives,
4. Make decisions based on available information and risk analyses,
5. Take actions to implement the decisions, and
6. Perform an evaluation of the effectiveness of the actions taken.

The first step of problem formulation corresponds to the first component of a sampling and monitoring program: *defining the purpose, questions, and objectives of the sampling and monitoring program*. At this step, ask if the correct questions were asked and if the data are of sufficient quantity and quality to answer the questions. If not, what are the data gaps or does more data need to be collected. Each case study in Appendix 5 discusses the purpose, questions, and objectives of the study.

The second step, analyzing the risks, was covered in Chapter 4. For the purposes of this handbook, *risk* is defined as the measure of the probability that damage to life, health, property, and/or the environment will occur as a result of a given hazard or condition.

For step three, defining the options or alternatives, it is recommended that the framework for decision making has the capacity to engage a comprehensive range of stakeholders and address the interdependence and cumulative effects of various options or alternatives.

Step four, making sound decisions, appears to be easy, but it is often the most difficult part of a sampling and monitoring program, especially if decision makers or the public were not involved early in the process of defining information needs. Sound environmental decision making is dependent on effective working relationships among scientists, policymakers, and the public based on reliable and timely *translations* of information and views between the communities. To facilitate communication between these groups, it is helpful to understand the general perspectives and potential biases of the groups. From the scientists' perspective, many management decisions are not made based on complete scientific data. From the policymakers' perspective, all scientists want to do is to collect more data. From the public's standpoint, risks that are voluntarily undertaken are downplayed. The public fears what it doesn't know and what it has no control over. The public tends to believe the first reports no matter the sources, even if they are preliminary and change with new data. Risks considered most serious by the general public can be different from those considered most serious by the technical professionals and policymakers charged with reducing environmental risk (see Education and Research Opportunities sidebar).

For example, some chemicals associated with the mining industry, such as cyanide, are not well understood by the public, have an automatic negative connotation in the media and public consciousness, and may need to be considered prominently in a sampling and monitoring program simply to improve public perception, understanding, and to allay concerns. The public process is doomed if the company is perceived to be reactive rather than proactive.

Step five, taking actions to implement the decisions, is considered self-explanatory. The action plan needs to be readily available to those implementing, monitoring, and analyzing the plan.

Step six is the step seldom taken in a sampling and monitoring program, but it is one of the most important steps: performing an evaluation of the appropriateness and effectiveness of the actions taken. Determine from all involved parties (the decision makers, stakeholders, and interested parties) if the information was adequate and useful. Documentation of lessons learned can provide amazing insights for your next sampling program. Each case study in Appendix 5 discusses the lessons learned from the selected study.

When in doubt regarding the final decision, one should use the precautionary principle as recommended by Morgan et al. (1990) as follows: "Increasingly, environmental decision-making

EDUCATION AND RESEARCH OPPORTUNITIES AT THE SLIPPERY ROCK WATERSHED, PENNSYLVANIA

Virginia T. McLemore, New Mexico Bureau of Geology and Mineral Resources

The Slippery Rock Watershed Coalition was created to restore a portion of the Slippery Rock Watershed in northwestern Pennsylvania that is impacted by acid mine drainage from abandoned coal mine complexes. The coalition has successfully utilized local college students from Grove City College in providing background data for passive treatment systems. In 1998, a scale model of a vertical flow system and an aerobic wetland was constructed in part by the college students to improve the quality of the water in that portion of the watershed. A monitoring program was developed and the students collected samples weekly as part of this program. Current results indicate that the system is viable for improving the water quality. This is one of many examples where the local community can be successfully involved in sampling and monitoring programs. In this case, the educational opportunities offered to the college students were invaluable (Brenner 2002), and these efforts lower the cost of the reclamation (Taylor et al. 2002).

A similar coalition was developed elsewhere in the Slippery Rock Watershed called the Jennings Water Quality Improvement Coalition. Elimination or abatement of degraded acid drainage from nearby coal mines has been attempted four times over the past 30 years, with little effect on the water quality. In 1997, the coalition, including government agencies, mining and environmental companies, Girl Scouts, college and home-school students, and other volunteers, constructed a vertical flow system. Monitoring of the system was by coalition volunteers from September 1997 through June 1999, and results indicate that the system continues to discharge alkaline water with low concentrations of metals (Taylor et al. 2002).

is scrutinized with respect to a precautionary principle. This principle asserts that where uncertainty and doubt make it impossible to be sure about a correct decision, any errors should favor the long-term sustainability of the environment" (Morgan and Henrion 1990; Raiffa 1997). The academic discipline of decision analysis under uncertainty, among others, has a rich literature on which to draw for methods and findings.

7.6.1 The Decision-Making Environment and the Importance of Process

The importance of attention to *process* is entirely compatible with the theory of the management sciences that defines a *good decision* under uncertainty as one that uses the most appropriate processes and methods to assemble and interpret evidence, to apply the decision makers' values properly, and to make timely choices with available resources, rather than defining a good decision only according to its (apparent) outcomes. Most decision-making situations involving matters of public health and environmental risk have four common elements:

1. The desire to use the best scientific methods and evidence to inform decisions
2. Uncertainty that limits the ability to characterize the magnitude of the problem and the corresponding benefits of proposed actions
3. The need for timely decision making that precludes resolving important uncertainties before the decisions are required
4. The presence of some sort of trade-off among adverse outcomes (which may be health, ecologic, or economic outcomes, each affecting a different set of stakeholders)

Most decisions that are based on data from sampling and monitoring programs are made to avoid risks, reduce risks, reduce the consequences of events, or ensure against the financial consequences of risks, as described in Chapter 4. Finally, the data, interpretation, and results need to be presented to the decision makers in a form that is useful to them such as in reports and presentations.

If monitoring data are limited, and when the decisions are under significant uncertainty, the precautionary principle comes into play. Where a course of action poses a risk that cannot be fully quantified, the precautionary principle provides conservative protection in the case of potential harm even if the probability cannot be determined accurately by the existing science (Crawford-Brown and Crawford-Brown 2011). Adaptive management refers to an iterative, systematic monitoring process to support robust decision making in the face of uncertainty. The U.S. Council on Environmental Quality issued a guidance memorandum regarding mitigation and monitoring that encourages the use of adaptive management (CEQ 2011):

> The guidance encourages agencies to develop internal processes for post-decision monitoring to ensure the implementation and effectiveness of the mitigation. It also states that agencies may use adaptive management as part of an agency's action. Adaptive management, when included in the NEPA [National Environmental Policy Act] analysis, allows for the agency to take alternate mitigation actions if mitigation commitments originally made in the NEPA and decision documents fail to achieve projected environmental outcomes.

A critical dimension of scope is the explicit inclusion of the various possible mitigation options that might be considered to reduce the risk that is being assessed. The scope may need to be expanded so that the assessment will provide not only estimates of existing risk, but estimates of risk reduction associated with changes in the risk-generating mining operation.

7.6.2 Decision Makers and Stakeholders

Interaction with the stakeholders (i.e., all interested parties), especially those making decisions, is an imperative, integral, and iterative part of a sampling and monitoring program. The public stakeholders should be involved throughout the project to maintain transparency and manage potential conflict caused by misconceptions or false data (see Stillwater Mining sidebar). The classic regulatory requirements of public participation should not be the goal, but the decision makers must be involved from the beginning of the process through to the end. Monitoring for monitoring's sake is worthless. Sampling and monitoring should be conducted to answer questions, and who better to define what is needed than those people asking the questions and making the decisions.

Legally required participation methods in decision making in the United States, such as public hearings, review and comment procedures in particular, are not always the best procedure. Not only do they not always meet most basic goals for public participation, but they can be counterproductive, causing anger, outrage, and mistrust. Research shows that collaborative participation can solve complex, contentious problems (Innes and Booher 2003). Participation in a sampling and monitoring program by the public can take on a variety of components. For example, local communities can have access to information and data unavailable to those conducting a monitoring program. They can also have questions that need to be answered as part of the sampling and analysis of data. Moreover, active participation in a sampling and monitoring effort tends to eliminate questions of data validity because participants have ownership in the process. This collaborative dialogue is a multi-way interaction in which citizens and other players work and talk in

STILLWATER MINING'S GOOD NEIGHBOR AGREEMENT

Carol C. Russell, U.S. Environmental Protection Agency

More than 10 years ago, Stillwater Mining Company, Northern Plains Resource Council, Stillwater Protective Association, and Cottonwood Resource Council established an innovative agreement that resolved disputes, protected the environment, and encouraged responsible economic development. The good neighbor agreement (GNA) provided solid criteria for the implementation of water monitoring plans, aggressive water discharge limits (that exceed normal regulatory standards), biological monitoring, reclamation, and closure plans with an emphasis on "best management practices" for all mining activities. A unique component of the GNA is that it provided the councils a voice in the decision-making process and an opportunity to review and comment on plans, procedures, and proposed solutions up front in the planning stage where issues can be addressed before development. This process minimizes the potential for conflicts that can slow or stall development and, at the same time, ensures that important environmental issues are addressed and factored into mining activities. The most recent environmental impact statement was processed in just a few months with the councils providing information to the regulatory agencies in support of the proposed amendment to the Stillwater long-term water management plans.

formal and informal ways to influence action in the public arena before it is virtually a foregone conclusion. The process is one of give and take, and joint problem solving, possibly in a social network cyberspace.

A common framework that allows for interacting and influencing one another and all are acting independently in the world results in a much better program. This is not one-way communication from citizens to government, or government to citizens, or company to government, or company to citizens. It is a multidimensional model where communication, learning, and action are joined together and where the policy, interests, and citizenry co-evolve. Innes and Booher (2003) describe such a model as follows:

> When an inclusive set of citizens can engage in authentic dialogue where all are equally empowered and informed and where they listen and are heard respectfully and when they are working on a task of interest to all, following their own agendas, everyone is changed. They learn new ideas and they often come to recognize that others' views are legitimate. They can work through issues and create shared meanings as well as the possibility of joint action.

The following tips are suggested for involving communities and stakeholders (USEPA 2000a):

- Know and respect the community.
- Establish an ongoing company presence.
- Maintain communication at all times.
- Devote special attention to historical preservation; concerns deserve special attention.
- Utilize the local community expertise.
- Involve the community in planning and implementing the sampling and monitoring program.

Mitigation and enhancement measures are best identified and developed in cooperation with local stakeholders. In 2013, the National Academy of Sciences (NAS 2013) suggested that "Early communication facilitates incorporation of stakeholder perspectives in the process and helps to identify uncertainties for consideration in the final decision." Furthermore, the International Council on Mining and Metals (ICMM 2010) produced *Good Practice Guidance on Health Impact Assessment in the Mining and Metals Sector*, an international reference on providing partnerships in decision making for local communities potentially affected by mining. When decisions must be made that affect groups of people or large segments of society, the process should be conducted within a structured framework that helps the decision makers take into account multiple objectives, needs, preferences, and values as well as vulnerabilities, risks, and uncertainties. The structured decision making is applied to break complex decisions into smaller, more tractable parts that are not prone to error and bias and are internally consistent.

7.7 SUMMARY

In order for data to be useful for decision making, it must be transformed into information by the process of assessment and analysis. Often the questions that decision makers would like to answer relate to compliance: what caused the problem, could the problem be prevented, what is the cost, and will it work? In addition, decision makers want to know the reliability of the data. Sometimes the scientific data is valid but the process used to make the decision is flawed. Ultimately, if data are not used to make decisions, they are of limited value.

7.8 INTERNET RESOURCES

ESRI's ArcGIS resources: http://resources.arcgis.com/content/web-based-help
R Project for statistical computing: www.r-project.org/
USEPA's STORET database: www.epa.gov/storet/

7.9 REFERENCES

Brenner, F.J. 2002. Public private partnerships: Education and research opportunities. SME Preprint 02-144. Littleton, CO: SME.

CEQ (U.S. Council on Environmental Quality). 2011. Memorandum on Appropriate Use of Mitigation and Monitoring and Clarifying the Appropriate Use of Mitigated Findings of No Significant Impact, dated January 14, 2011. Washington, DC: CEQ.

Cormier, S.M., and Suter, G.W. 2008 A framework for fully integrating environmental assessment. *Environ. Manage.* 42:543–556.

Crawford-Brown, D., and Crawford-Brown, S. 2011. The precautionary principle in environmental regulations for drinking water. *Environ. Sci. Policy* 14:370–387.

Grunsky, E.C. 1997. Strategies and methods for the interpretation of geochemical data. In *Current Topics in GIS and Integration of Exploration Datasets*. Short Course, Exploration '97 Workshop. Ottawa, ON: Geological Survey of Canada.

Helsel, D.R., and Hirsch, R.M. 2002. *Statistical Methods in Water Resources*. Techniques of Water Resources Investigations, Book 4. Reston, VA: U.S. Geological Survey. http://pubs.usgs.gov/twri/twri4a3/. Accessed 9/11/13.

ICMM (International Council on Mining and Metals). 2010. *Good Practice Guidance on Health Impact Assessment in the Mining and Metals Sector*. ICMM report. www.icmm.org. Accessed 9/11/13.

Iglewicz, B., and Hoaglin, D.C. 1993. *How to Detect and Handle Outliers*. Milwaukee, WI: American Society for Quality Control.

Innes, J.E., and Booher, D.E. 2003. Collaborative policymaking through dialogue. In *Deliberative Policy Analysis: Understanding Governance in the Network Society*. Edited by M.A. Hajer and H. Wagenaar. New York: Cambridge University Press.

Isaaks, E.H., and Srivastava, R.M. 1989. *An Introduction to Applied Geostatistics*. New York: Oxford University Press.

McLemore, V.T. 2010. Use of the New Mexico Mines Database and ArcMap in uranium reclamation studies. SME Preprint 10-125. Littleton, CO: SME. http://geoinfo.nmt.edu/staff/mclemore/documents/11-139.pdf. Accessed 9/11/13.

McLemore, V.T., Hoffman, G.K., Wilks, M., Raugust, J.S., and Jones, G.R. 2004. Use of databases in characterization at mine sites. In *Proceedings, 2004 National Meeting of the American Society of Mining and Reclamation*. Lexington, KY: American Society of Mining and Reclamation.

Morgan, G.M., and Henrion, M.H. 1990. *Uncertainty: A Guide to Dealing with Uncertainty in Quantitative Risk and Policy Analysis*. New York: Cambridge University Press.

NAS (National Academy of Sciences). 2013. *Environmental Decisions in the Face of Uncertainty*: Washington D.C. National Academies Press. www.nap.edu/catalog.php?record_id=12568. Accessed 9/11/13.

NRC (National Research Council). 2001. *Assessing the TMDL Approach to Water Quality Management*. Washington, DC: National Academies Press.

Raiffa, H. 1997. *Decision Analysis: Introductory Readings on Choices Under Uncertainty*. New York: McGraw-Hill.

Rollinson, H.R. 1993. *Using Geochemical Data: Evaluation, Presentation, Interpretation*. Harlow, England: Pearson Education, Prentice Hall, Longman Group.

Schreuder, H.T., Ernst, R., and Ramirez-Maldonado, H. 2004. *Statistical Techniques for Sampling and Monitoring Natural Resources*. General Technical Report-RMRS-GTR-126. Fort Collins, CO: U.S. Department of Agriculture, Rocky Mountain Research Station. www.fs.fed.us/rm/pubs/rmrs_gtr126.pdf. Accessed 9/11/13.

Stewart, T.R. 2000. Uncertainty, judgment, and error in prediction. In *Prediction: Science, Decision Making, and the Future of Nature*. Edited by D. Sarewitz, R.A. Pielke, and R. Byerly. Washington, DC: Island Press. pp. 41–57.

Taylor, W., Johnson, D., Dunn, M., and Hedin, R. 2002. A public-private partnership success story: Treating abandoned mine drainage at Jennings Environmental Education Center. SME Preprint 02-144. Littleton, CO: SME.

USEPA (U.S. Environmental Protection Agency). 2000a. *Abandoned Mine Site Characterization and Cleanup Handbook*. U.S. Environmental Protection Agency. EPA /910-B-00/-001. www.epa.gov/superfund/policy/remedy/pdfs/amscch.pdf. Accessed 9/11/13.

USEPA (U.S. Environmental Protection Agency). 2000b. *Guidance for Data Quality Assessment—Practical Methods for Data Analysis* (EPA QA/G-9 QA00 update). U.S. Environmental Protection Agency. EPA/600/R-96/084. www.epa.gov/region6/qa/qadevtools/mod4references/secondaryguidance/g9-final.pdf. Accessed 9/11/13.

USEPA (U.S. Environmental Protection Agency). 2002. *Guidance on Environmental Data Verification and Data Validation*: U.S. Environmental Protection Agency. EPA QA/G-8 240/R-02/004. www.epa.gov/QUALITY/qs-docs/g8-final.pdf. Accessed 9/11/13.

USEPA (U.S. Environmental Protection Agency). 2006. *Data Quality Assessment: Statistical Methods for Practitioners* (EPA QA/G-9S). U.S. Environmental Protection Agency. EPA/240/B-06/003. www.epa.gov/QUALITY/qs-docs/g9s-final.pdf. Accessed 9/11/13.

USEPA (U.S. Environmental Protection Agency). 2008. *Scout 2008 Version 1.0–Quick Overview: A Statistical Software Package for the Classical and Robust Analysis of Univariate and Multivariate Data Sets with and Without Non-Detect Observations*: U.S. Environmental Protection Agency. EPA/600/F-08/016. www.epa.gov/esd/databases/scout/Scout%202008%20v.1.00.01/Scout%20Resources/Scout%202008%20Guides/Quick%20Overview%20Scout%202008%20v.%201.0.pdf. Accessed 9/11/13.

USEPA (U.S. Environmental Protection Agency). 2010. *Integrating Ecological Assessment and Decision-Making at EPA: A Path Forward Results of a Colloquium in Response to Science Advisory Board and National Research Council Recommendations*. EPA/100/R-10/004. www.epa.gov/raf/publications/pdfs/integrating-ecolog-assess-decision-making.pdf. Accessed 9/11/13.

USEPA (U.S. Environmental Protection Agency). 2011. STORET database. www.epa.gov/storet/. Accessed 9/11/13.

USEPA (U.S. Environmental Protection Agency). 2013. Metadata: Water Quality Observation. http://water.epa.gov/scitech/datait/models/basins/wqobs.cfm. Accessed 12/20/13.

Weight, W.D., and Sonderegger, J.L. 2001. *Manual of Applied Field Hydrogeology*. New York: McGraw-Hill.

CHAPTER 8

Additional Key Issues and Future Research Needs

Virginia T. McLemore

8.1 INTRODUCTION

Structural problems associated with mines, process facilities, mine structures, and waste management units must be considered to ensure the safety of workers and alleviate environmental impacts that could result from structural failure and a possible release of contaminants into the local drainage. Even avalanches could affect the best sampling and monitoring plans (see The Effect of Avalanches sidebar). Most of these features should have a monitoring plan to examine them periodically during the mine life cycle and even after operations. The following sections briefly describe some of these stability issues and their effect on water quality. This chapter concludes with some comments on research needs and topics.

8.2 SLOPE STABILITY AND FAILURE

A slope failure is the uncontrolled or unscheduled release of the slope material beyond the confines of the original topography (Robertson and Skermer 1988; Neuendorf et al. 2005). Most slope failures are caused by disruptive actions that can be described in one of two ways:

1. Sudden, intense, or extreme events such as floods that cause liquefaction (Hutchinson 1988), earthquakes, volcanic action, avalanches, and glaciation, which apply forces exceeding the values for which the slopes were originally designed.
2. Slow, but perpetual actions of wind and water erosion, including frost action, lubrication, other forms of weathering and decomposition, chemical reaction, and biological actions such as intrusion by roots, animals, and humans.

The most common slope failures at mine sites are from slopes forming the open pit walls, mine waste piles, and process tailings impoundments, although any slope that has been disturbed by mining or construction activities can fail under the right conditions, including slopes adjacent to roads and railroads.

Two types of slopes are associated with mine sites: cut slopes and manufactured or filled slopes. The excavation of overburden and/or ore creates cut slopes, also called benches or highwalls. Manufactured or filled slopes are formed by the dumping or piling of overburden, process tailings, waste rock, or other materials; these features can be high in metals, acid forming, or reactive. Slope failure as a result of gravitational forces or by saturation of the material by water can result in direct release or direct exposure of these materials to the surrounding weathering environment. Changes to an existing slope can create environmental problems, such as increased erosion, rapid runoff, sedimentation, changes in wildlife patterns, and the exposure of potentially

THE EFFECT OF AVALANCHES AT THE OPHIR MINE, COLORADO

Carol C. Russell, U.S. Environmental Protection Agency

Ophir, a small mining town located 12 km (7.5 miles) from Telluride, Colorado, was heavily damaged by snow avalanches in the early 1900s during the mining boom. Active treatment of acid rock drainage (ARD) in this area would be very expensive and would be hampered by liability provisions of the Clean Water Act. Likewise, more passive-treatment strategies, such as diverting ARD through constructed wetlands and bioreactors, must still be regarded as experimental because of the cold temperatures at this elevation. In addition, the site is potentially hazardous because of the risk of destruction by snow-slides. The U.S. Environmental Protection Agency recommends strategies to intercept and divert water away from mineralized zones, thereby reducing metal loading to the Howard Fork.

reactive natural materials. In many instances, slope failures result in degradation of the drainage quality of the receiving stream.

Slope failures and breaches in tailings piles are more common, whereas sudden slope failures of mine rock piles are not as common. However, when any slope failure occurs, the resulting publicity feeds public perceptions that these mine features are hazardous.

Four types of sudden, intense failures of mine-waste rock piles are (Pernichele and Kahle 1971; Whiting 1981)

1. Debris flow (also known as flow slide),
2. Foundation failure,
3. Edge slump, and
4. Blowout (internally generated by elevated hydraulic pressures that accumulate in some layers resulting in decreasing internal effective stress and stability).

Many investigators have studied the mechanisms that result in failed rock piles, which are summarized by McLemore et al. (2009). Any disturbances should be routinely sampled and monitored throughout the mine life cycle.

8.3 TAILINGS IMPOUNDMENTS

Tailings are the fine-grained residue of mineral processing typically found at mill sites. Some other terms applied to tailings include slimes, tails, leach residue, or slickens. The volume of tailings generally exceeds the volume of the recovered minerals by several orders of magnitude. Most tailings contain water and are stored in impoundments, typically called tailings storage facilities (or ponds), where the water is reused in the mill or allowed to evaporate and the solid particles settle to the bottom. The tailings are contained by liners, natural rock formations, and human-made steep slopes called tailings dams and generally consist predominantly of sand-size and smaller fragments.

Selected types of tailings release and damage to the environment at tailings storage facilities include the following (USCOLD 1994; Klohn 1995; Davies 2002; Blight and Fourie 2003; Mayoral and Romo 2008; Azam and Li 2010):

- Unusual weather
- Overtopping

- Static instability or structural defect (liner failure)
- Seismic instability (earthquakes)
- Internal erosion or piping
- Wind-blown dust
- Seepage or drainage of liquid wastes, such as acid drainage
- Mine subsidence
- Weak foundation
- Excessive water level rise

Not only can water quality be affected, but the entire downstream ecosystem can be adversely impacted. Flooding can contaminate surface water, groundwater, and soil, and it can adversely affect wildlife and humans.

Although engineering of tailings dams has improved since the 1990s, tailings dam failures still occur nearly every year somewhere in the world (Table 8.1; Davies et al. 2002; McLemore et al. 2009; Azam and Li 2010) and cost millions of dollars to remediate. These tailings dam failures can have serious impacts downstream. In Europe, the most common cause of tailings dam failures is related to unexpected rainfall above the engineered strength of the tailings dam (Rico et al. 2008). Elsewhere in the world, other causes of tailings dam failure are seismic liquefaction (Rico et al. 2008), inadequate management, errors in site selection, poor foundation, and lack of stability of downstream slopes. Most of the dam failings occur at active mine sites (Davies et al. 2002; Azam and Li 2010), where sampling and monitoring plans are generally in place. Sampling and monitoring of water within and surrounding tailings facilities and reclaimed water are typically routine.

The design, construction, operation, and decommissioning of tailings dams must be carefully planned in order to ensure their safe performance. Safe tailings dams can be built by utilizing sound engineering practices and understanding climatic and tectonic conditions. Lessons learned from previous incidents should be incorporated into any future tailings design.

8.4 SUBSIDENCE

Mine subsidence is the collapse of overlying geologic formations into mine voids, and the potential for subsidence exists for all underground mining. Effects of subsidence are not always visible at the ground surface. Subsidence can manifest itself in the form of sinkholes or troughs at the surface. Sinkholes are usually formed by the collapse of a portion of a mine void (such as a room in room-and-pillar mining or block caving techniques). The extent of the surface disturbance is usually limited in size. Natural sinkholes and caves are common in terrains underlain by limestone or other carbonate or evaporate rock, and mining can enlarge these natural features. Many types of underground mining can result in subsidence at the surface. Troughs are formed from the subsidence of large portions of the underground void and typically occur over areas where most of the resource has been removed. Sinkholes or depressions in the landscape interrupt surface-water drainage patterns; ponds and streams may be drained or channels can be redirected. In developed areas, subsidence has the potential to affect cultural features such as building foundations and walls, highways, and pipelines. Subsidence can contribute to increased infiltration to underground mines, potentially resulting in increased acid rock drainage (ARD) generation. Groundwater flow can be interrupted as impermeable strata break down and could result in flooding of the mine and drying up of water wells. Impacts to groundwater include changes in water quality, permeability, and drainage flow patterns (including surface-water recharge). Monitoring

TABLE 8.1 Recent tailings dam incidents

Date	Location	Company	Type of Ore	Description	Release	Impacts	Reference
July 21, 2011	Mianyang City, Songpan County, Sichuan Province, China	Xichuan Minjiang Electrolytic Manganese Plant	Mn	Tailings dam damaged from landslides caused from heavy rains	?	Damaged roads, houses, and local water supply	WISE (2013)
October 4, 2010	Kolontár, Hungary	MAL Magyar, Aluminum	Bauxite	Tailings dam failed	700,000 m^3 of red mud	Several towns (8 km^2) flooded; 10 people killed	Boston.com (2010)
June 25, 2010	Huancavelica, Peru	United Minera Caudalosa Chica	Ag, Cu, Pb, Zn ore	Tailings dam failed	21,420 m^3 of tailings	Contamination of Rio Escalera and Rio Opamyo 110 km downstream	Guerra (2010)
August 29, 2009	Karamken, Russia	—	Gold	Tailings dam failed after heavy rain	—	11 homes destroyed; 1 person killed	WISE (2013)
December 22, 2008	Kingston Fossil Plant, Harriman, Tennessee	—	Coal ash	Retention wall failed	4.1 m^3	Covered 400 acres; damaged 12 homes	Dewan (2008)

NOTE: Most of these failures occurred outside of the United States. Other incidents are summarized at www.tailings.info/accidents.htm.

of the surface topography throughout the mine life cycle and even after closure can predict subsidence issues at most facilities.

8.5 MINE OPENINGS

Horizontal and vertical mine openings can be physical hazards and potential threats to the general population if not properly secured. Adventurous individuals want to enter them to explore, and can possibly be injured or killed as a result of unstable ground or bad air (e.g., insufficient oxygen or presence of poisonous gases such as carbon monoxide). Mine openings can also be the source of drainage of mining influenced water (MIW) that eventually enters the local drainage system, possibly causing contamination. Draining adits or tunnels are a particular concern, because these adits were excavated beneath the ore horizons with the sole purpose of draining the ore body and typically contain a variety of metals, which can enter the local drainage system. Detailed sampling and monitoring of these features can determine if they pose an environmental threat. The Argo Tunnel near Idaho Springs, Colorado (Figure 8.1)—an adit used to drain upper mine workings—typically produces water flow rates ranging from 200 to 300 gpm, depending on the season (see sidebar in Chapter 2). The pH ranges between 2.7 and 3.2, and the water contains high concentrations of zinc, iron, sulfate, cadmium, copper, and manganese.

8.6 CLIMATE CHANGE

Climate changes are natural processes and found throughout the geologic past. It is not as important to identify the source of potential climatic changes (i.e., emissions from humans, volcanic eruptions, deep sea methane release, etc.) as it is to recognize that climate changes are inevitable. Potential changes include, but are not limited to, the following (Trenberth 1999; Arctic North Consulting 2009; Geyer and Wasilewski 2009):

Courtesy of Virginia T. McLemore.

FIGURE 8.1 Argo Tunnel near Idaho Springs, Colorado

- Temperature changes, over the course of a year and during the seasons
- Changes in precipitation and in relative amounts of snowfall versus rainfall
- Changes in timing of precipitation events
- Rising sea level due to melting of ice caps
- Air quality changes
- Increases in wind, floods, heat waves, and storms (cyclones, tornadoes, hurricanes, ice storms)

Future climate changes and extremes can impact the mining industry in different ways (OCCIAR, n.d.; Nelson and Schuchard, n.d.). Two major effects of climate change should be considered: how the mine operations may impact the future climate and how the mine company will adapt to a changing climate. A changing climate can result in modifications of existing sampling and monitoring plans.

Very large mines and process facilities have the potential to impact local and even global climate change by increasing the amount of carbon dioxide (CO_2) due to acid neutralization by limestone, kiln and roaster process operations, and other greenhouse gases in the atmosphere. Most operations in the United States will be required to evaluate the carbon impact from their operations. As forests and other vegetation are cleared, CO_2 is no longer taken up by the vegetation and remains in the atmosphere. CO_2 and other gases can be released into the atmosphere by mining equipment and processing facilities (Norgate and Rankin 2000) or through liberation of methane from coal and other sedimentary units during mining. Environmental Law Alliance Worldwide (2010) estimates that more than 1 kg of greenhouse gas is emitted into the atmosphere for every 1 kg of metal produced (Norgate and Rankin 2000). Studies of climates in geologic times suggest that an increase in CO_2 levels can result in a more humid climate in some areas that are higher in precipitation and temperature (Retallack 2009). Other areas can experience drier conditions. Some mining companies are taking steps to improve the energy efficiency of their mining and processing operations. More research is needed to understand what, if any, effects mining and processing can have on climate and how to mitigate such effects.

Mining companies that plan to accommodate changes in climate, including the proposed CO_2 regulations, and identify potential benefits that can be achieved could be perceived as being proactive and be received more favorably by the local communities. Mining companies will also be the source for raw materials needed to produce alternative forms of energy (e.g., solar panels,

USE OF BIOCHAR TO REDUCE ACID ROCK DRAINAGE AND STORE CO_2

Virginia T. McLemore, New Mexico Bureau of Geology and Mineral Resources

Biochar is charcoal created by pyrolysis (heating plant and other organic material in the absence of oxygen) of biomass and can be used as a soil amendment and fertilizer at mine sites. The biomass can be dead wood from forest fires or insect damage, or wastes from poultry farms. Biochar adds carbon and other nutrients to the soil, improves the soil's ability to hold water, and is suited for low-pH, low-humus soils. Biochar added to mine wastes can reduce acid drainage, improve water quality, and increase plant growth. In addition, biochar may actually store CO_2, N_2O, and CH_4 (i.e., greenhouse gases). Field tests are under way at mine sites in the Rocky Mountains and in West Virginia to determine the effectiveness of biochar as a soil amendment in mine reclamation (West Virginia University 2011; IBI 2013; Harley 2010).

wind turbines, electric/hybrid cars). Nuclear power, which requires the mining of uranium, could be a larger component of the local and national power supply in the future. Process tailings facilities in New Mexico and Arizona are currently the sites of large solar panel arrays that supply electricity to the surrounding communities, thereby providing for a sustainable postmining use. Mine wastes and process tailings could be potentially effective in absorbing CO_2 and storing carbon; the use of biochar is one example (see Biochar sidebar; Harley 2010). As research progresses, other benefits are likely to be realized.

The effects of changing climate are being researched throughout the world (Lemmen et al. 2008; Arctic North Consulting 2009; Geyer and Wasilewski 2009; Nelson and Schuchard, n.d.). However, mining operations today are designed in respect to historical climate changes instead of projected future climate changes.

Mines may see changes in both weather and climate (Nelson and Schuchard, n.d.). Some areas could see more frequent weather changes than the mine is now facing or even new changes in the weather, such as excessive rainfall. Excessive rainfall could lead to increases and decreases in ARD and metals concentrations in waters, in both MIW and waters from adjacent mineralized but unmined areas. Sedimentation could increase as rainfall increases. After prolonged droughts, sudden increases in ARD and metals concentrations would occur during the early part of storm events, sometimes called *first flush* (Nordstrom 2008). However, as weather changes in other areas, mines could face events not always foreseen nor engineered to withstand, such as flooding, erosion (both wind and water), landslides, heat waves, overflowing of impacted water containment ponds, debris flows, increased snowfall and freezing rain, avalanches, increased wildfires, and amplified winds that can result in increased dust emissions. Landslides, mudslides, and flooding can occur after wildfires. Changes can be dramatic even between adjacent water basins (Hanson et al. 2006). Increased temperatures could result in increased evaporation from tailings facilities. Heat waves could stress workers and equipment, possibly resulting in production delays. Hotter and drier climates can increase wildfires that could threaten mines and processing sites. Existing mining infrastructures are not always designed for extreme changes in weather conditions. All of these effects could impact water quality. Water supplies could be affected by short-term weather and long-term climate changes. New technologies are needed to increase reuse and reduce water use in mines and processing facilities. Also, weather and ultimately climate change could affect migration of animals. All of these factors should be considered in the mine and processing plans; the mine plans should at least be flexible enough to incorporate changes if these factors play a role.

8.7 SAMPLING AND MONITORING IN OTHER COUNTRIES

Water quality impacts by mining activities is a major concern in most countries in the world and has resulted in various regulatory programs by government agencies. Sampling and monitoring plans, although not always specifically required by the regulatory agencies throughout the world, are fundamental to achieving desired results. Most international mining companies use similar sampling and monitoring programs in their mines and processing facilities regardless of their location, which results in environmental programs exceeding national requirements at many international mines.

Stakeholder engagement and respect for indigenous cultures are becoming increasingly important throughout the world (Mineral Resources Group 2009). International organizations such as the World Bank, International Finance Corporation, and World Health Organization can provide additional guidance.

A description of the regulatory process and sampling and monitoring plans throughout the world is beyond the scope of this handbook. However, a summary of the regulatory requirements for the major countries can be found in the *Global Acid Rock Drainage Guide* (INAP 2009). Some common themes worldwide include the following (INAP 2009):

- Environmental management plans are typically formed as part of the authorization process and can describe how the effects will be managed during the mining operational phase.
- The prediction of conditions and planning for mine closure should be determined, which could require numerous sampling and monitoring plans.
- Ongoing monitoring, review, and continuous improvement, particularly for ARD, are recommended.

Many areas of the world have summarized requirements for mine reclamation and avoiding water quality risks. Australia recently enacted new regulations. Price and Errington (1998) and Mine Environment Neutral Drainage (MEND 2005) provide a comprehensive list of the information requirements and factors to consider regarding ARD management in mines operating in Canada, and Campbell (2004) discusses regulations in Africa. The European Commission is responsible for developing policies and proposals for regulations of mines, under which the European Union member states will operate (INAP 2009). In Australia, a mining and rehabilitation program (MARP) must be prepared that documents risks and provides specific and detailed information on management and control measures for those risks (Mineral Resources Group 2009, 2010). The MARP includes a complete description of the mine site and planned operations, documentation of consultation with communities and indigenous people, description and documentation of risks, and management of those risks.

8.8 FUTURE RESEARCH NEEDS AND TOPICS

As in most studies, more questions are realized:

- How do we better measure quantitative mineralogy?
- How can supergene minerals originally in the mined rock be differentiated from secondary minerals originating from weathering in a mine-waste storage facility?
- What are the impacts from ARD or MIW from industrial minerals mines? (An important step would be to perform a literature search of existing data to identify sources of information needed to quantify impacts on water from industrial minerals mines in the United States in terms of stream miles, watershed acreage, number of sites, and groundwater acreage. To accomplish this, a consensus-reviewed report could be produced that can be used

as the basis for a request for proposal to carry out the study on the impacts of historical mined lands on water resources.)

- What are the risks of mine storage facilities to produce ARD? (Sampling and testing techniques need to be further developed to address the hydrologic and geochemical characterization of mine storage facilities.)
- What do differences in stable isotopes in different types of waters mean, and can isotopes characterize mixing and mechanisms of mixing?
- What types, formats, content, and inventories of database management are needed, and how should they be managed?
- What is the effect of MIW from sand and gravel/aggregate pits?
- What is the geochemistry/mineralogy of cast-off products from industrial activities and how do they affect MIW (i.e., cement, dredging, mine wastes, organic wastes, fines)?
- How do MIW and ARD affect sediments, sediment transport, and contamination?
- How do mine and forest fires affect water quality at mine and processing sites?
- How do blowouts affect water quality?
- How do we deal with advancements in technologies, new technologies, and integration of technologies involving water quality at mine sites?

In addition, the following research items reflect a future need:

- Development of improved selective/partial extraction techniques to identify and quantify mineralogy and its effect on water quality.
- Better methods for the long-term monitoring of remediation sites.
- Case studies of remediation and models of drainage quality at mine sites.
- Development of improved methods to model mine hydrology.
- Use of stable and radiometric isotopic data in drainage systems, including groundwater.

8.9 REFERENCES

Arctic North Consulting. 2009. *Climate Change and Canadian Mining: Opportunities for Adaptation.* Report prepared for the David Suzuki Foundation. www.davidsuzuki.org/publications/downloads/2009/Climate_Change_And_Canadian_Mining.pdf. Accessed 8/23/13.

Azam, S., and Li, Q. 2010. *Tailings Dam Failures: A Review of the Last One Hundred Years.* Waste Geotechnics. www.infomine.com/publications/docs/Azam2010.pdf. Accessed 8/23/13.

Blight, G.E., and Fourie, A.B. 2003. *A Review of Catastrophic Flow Failures of Deposits of Mine Waste and Municipal Refuse.* Report. ww.unina2.it/flows2003/flows2003/articoli/G.E.%20BLIGHT%20&%20A.B.%20FOURIE.pdf. Accessed 11/20/13.

Boston.com. 2010. The big picture: A flood of toxic sludge. October 6. www.boston.com/bigpicture/2010/10/a_flood_of_toxic_sludge.html. Accessed 9/16/13.

Campbell, B., ed. 2004. Regulating mining in Africa: For whose benefit? Nordiska Afrikainstitutet, Discussion Paper 26. http://nai.diva-portal.org/smash/record.jsf?searchId=1&pid=diva2:240515. Accessed 8/23/13.

Davies, M.P. 2002. *Tailings Impoundment Failures: Are Geotechnical Engineers Listening?* Infomine report. www.infomine.com/publications/docs/Davies2002.pdf. Accessed 8/23/13.

Davies, M., Martin, T., and Lighthall, P. 2002. *Mine Tailings Dams: When Things Go Wrong.* Infomine report. www.infomine.com/publications/docs/Davies2002d.pdf. Accessed 8/23/13.

Dewan, S. 2008. Tennessee ash flood larger than initial estimate. *New York Times.* December 26. www.nytimes.com/2008/12/27/us/27sludge.html. Accessed 9/16/13.

Environmental Law Alliance Worldwide. 2010. *Guidebook for Evaluating Mining Project EIAs*. Eugene, OR: Environmental Law Alliance Worldwide. www.elaw.org/files/mining-eia-guidebook/Full-Guidebook.pdf. Accessed 8/23/13.

Geyer, T., and Wasilewski, C. 2009. Climate change risk and impact assessment for global diversified mining group. SME Preprint 09 015. Littleton, CO: SME.

Guerra, I. 2010. Tailings dam collapses in Huancavelica, Peru. *Mines and Communities*. July 4. www.minesandcommunities.org/article.php?a=10212. Accessed 9/16/13.

Hanson, R.T., Dettinger, M.D., and Newhouse, M.W. 2006. Relations between climatic variability and hydrologic time series from four alluvial basins across the southwestern United States. *Hydrogeol. J.* 14:1122–1146.

Harley, A.D. 2010. Biochar activities in the Rocky Mountains with respect to mine reclamation (abstract). Geological Society of America, *Abstr. Prog.* 42(5):158.

Hutchinson, J.N. 1988. General Report. Morphological and geotechnical parameters of landslides in relation to geology and hydrogeology. 5th International Symposium on Landslides, Switzerland, pp. 3–35.

IBI (International Biochar Initiative). 2013. Learn more about biochar. www.biochar-international.org/. Accessed 9/16/13.

INAP (International Network for Acid Prevention). 2009. *Global Acid Rock Drainage Guide* (GARD Guide). www.gardguide.com/index.php/Main_Page. Accessed 8/23/13.

Klohn, E.J. 1995. Tailings dams: Their evolution to major hydraulic structures and the importance of their safety. Keynote address to Canadian Dam Safety Association Conference, Banff, Alberta, October.

Lemmen, D.S., Warren, F.J., Lacroix, J., and Bush, E., eds. 2008. *From Impacts to Adaptation: Canada in a Changing Climate 2007*. Ottawa, ON: Government of Canada. www.nrcan.gc.ca/earth-sciences/climate-change/community-adaptation/assessments/132. Accessed on 9/16/13.

Mayoral, J.M., and Romo, M.P. 2008. Geo-science environmental aspects affecting tailings dams failures. *Am. J. Environ. Sci.* 4:212–222.

McLemore, V.T., Fakhimi, A., van Zyl, D., Ayakwah, G.F., Boakye, K., Anim, K., Ennin, F., Felli, P., Fredlund, D., Gutierrez, L.A.F., Nunoo, S., Tachie-Menson, S., and Viterbo, V.C. 2009. *Literature Review of Other Rock Piles: Characterization, Weathering, and Stability*. Open-File Report 517. New Mexico Bureau of Geology and Mineral Resources. http://geoinfo.nmt.edu/publications/openfile/details.cfml?Volume=517. Accessed 9/16/13

MEND (Mine Environment Neutral Drainage). 2005. *List of Potential Information Requirements in Metal Leaching/Acid Rock Drainage (ML/ARD) Assessment and Mitigation Work*. MEND Report 5.10E. www.mend-nedem.org/reports/files/5.10E.pdf. Accessed 9/16/13.

Mineral Resources Group. 2009. *Guidelines for Miners: Mining Approval Processes in South Australia*. Government of South Australia, Minerals Regulatory Guidelines MG1, version 1.9. www.pir.sa.gov.au/__data/assets/pdf_file/0011/94952/mg01_mining_approvals_v1.9.pdf.

Mineral Resources Group. 2010. *Guidelines for Miners: Preparation of a Mining and Rehabilitation Program (MARP) for Extractive Mineral Operations in South Australia*. Government of South Australia. Minerals Regulatory Guidelines MG1, version 1.9. www.pir.sa.gov.au/__data/assets/pdf_file/0006/71673/MG06_prepn_marp_extractives_v1.9.pdf. Accessed 9/16/13.

Nelson, J., and Schuchard, R. n.d. Adapting to climate change: A guide for the mining industry. BSR industry series. www.bsr.org/reports/BSR_Climate_Adaptation_Issue_Brief_Mining.pdf. Accessed 8/20/13.

Neuendorf, K.K.E., Mehl, J.P., Jr., and Jackson, J.A. 2005. *Glossary of Geology*, 5th ed. Alexandria, VA: American Geological Institute.

Nordstrom, D.K. 2008. Acid rock drainage and climate change. *J. Geochem. Explor.* 100:97–104.

Norgate, T.E., and Rankin W.J. 2000. Life cycle assessment of copper and nickel production. In *Proceedings, Minprex 2000, International Conference on Minerals Processing and Extractive Metallurgy*. pp. 133–138. www.minerals.csiro.au/sd/CSIRO_Paper_LCA_CuNi.htm. Accessed 9/16/13.

OCCIAR (Ontario Centre for Climate Impacts and Adaptation Resources). n.d. Mining: In a changing climate. Fact sheet. www.climateontario.ca/doc/factsheets/Mining%20Factsheet%20--%20Final.pdf. Accessed 8/20/13.

Pernichele, A.D., and Kahle, M.B. 1971. Stability of waste dumps at Kennecott's Bingham Canyon mine. *AIME Trans.* 250:363–367.

Price, W.A., and J.C. Errington. 1998. *Guidelines for Metal Leaching and Acid Rock Drainage at Mine Sites in British Columbia.* Ministry of Energy and Mines, August.

Retallack, G.J. 2009. Greenhouse crises of the past 300 million years: *Geolog. Soc. Am. Bull.* 121:1441–1445.

Rico, M., Benito, G., Salgueiro, A.R., Diez-Herrero, A., and Pereira, H.G. 2008. Reported tailings dam failures: A review of the European incidents in the worldwide context. *J. Hazard. Mater.* 152:846–852.

Robertson, A., and Skermer, N.A. 1988. Design considerations for the long term stability of mine wastes. First International Environmental Workshop, Darwin.

Trenberth, K.E. 1999. Conceptual framework for changes of extremes of the hydrological cycle with climate change. *Climate Change* 42:327–339.

USCOLD (U.S. National Committee on Large Dams). 1994. *Tailings Dam Incidents.* Report prepared by the USCOLD Committee on Tailings Dams, November.

West Virginia University. 2011. Research highlights. http://research.wvu.edu/wvu_research_highlights/2011/7/25/from-farm-to-fuel-to-fields. Accessed 7/9/11.

Whiting, D.L. 1981. Surface and groundwater pollution potential. In M.K. McCarter, ed. *Design of Non-impounding Mine Waste Dumps.* New York: AIME. pp. 90–98.

WISE (World Information Service on Energy). 2013. Uranium project: Chronology of major tailings dam failures. www.wise-uranium.org/mdaf.html. Accessed 9/16/13.

Index

NOTE: *f.* indicates figure; *t.* indicates table.

A

Abandoned Mine Site Characterization and Cleanup Handbook, 127, 155
Abandoned mines. *See* Legacy mines
Access (software), 154
Accuracy, 81
  defined, 80, 80*f.*
Acid Drainage Technology Inititative–Coal Mining Sector, 2–3
Acid mine drainage (AMD), 2–3, 4–5
  and mineralogical studies, 20
  related terminology, 127
Acid rock drainage (ARD), 4–5
  international management regulations, 177
  quality and certainty of data, 43
  related terminology, 127
  research needs and topics, 177–178
  and sulfide-bearing rock, 10–12
  testing for potential in predevelopment environmental assessment, 52
Acid-producing rock (APR), 70
Acidic drainage, 1, 4–5
  caused during exploration, 10–12
  found during construction of tailings dam, 12
  and legacy mines, 27
  from lowering of water table, 22
  at prospect sites (pyrite), 12–13
  related terminology, 127
  and sulfide-bearing rock, 10–12
Acoustic Doppler current profilers (ADCPs), 47–49
Acoustic Doppler velocimeters, 47
Adaptive management, 167
Agricola, Georgius, on risk in mining, 113
Air rotary drilling, 72
Air rotary reverse circulation drilling, 72
Air sampling, 76–78
  and blasting, 78, 79*f.*
  and National Environmental Policy Act, 77
  and personal exposure sampling, 78
  targeted monitoring, 77
  and various emissions throughout mine life cycle, 77
AMD. *See* Acid mine drainage
Analytical chemistry, 84
  and cost-effectiveness, 86
  and interferences, 84
  laboratory selection, 86
  and laboratory subsampling, 85–86
  and matrix effects, 85
  method selection, 84
  references, 84
  and sample preparation, 85–86
  and volatility of constituents, 85
APR. *See* Acid-producing rock
ArcGIS software, 164
ARD. *See* Acid rock drainage
Argo Tunnel (Colorado), 175*f.*
  and data quality and certainty, 43
  and mining influenced water, 174
Assessment, scientific, defined, 155
Auger sampling, 72
Australia
  mining and rehabilitation programs (MARPs), 177
  sampling and monitoring guidance, 122*f.*, 123, 124*t.*
Automated SEM-based quantitative mineralogical analysis, 60

B

Background
  ambient, defined, 36

defined, and distinguished from baseline, 36
natural, defined, 36
Bar charts, 158
Baseline, defined, and distinguished from background, 36
Baseline-monitoring programs, 5
Bias, 39, 67, 68*f.*, 80–81
causes of, 84
defined, 79, 80, 80*f.*
Bioaccessibility tests, 61
Bioaccumulation factors (BAFs), 117
Bioavailability, and toxicity, 117, 117*f.*
Bioconcentration factors (BCFs), 117
Biological sampling (aquatic), 73–74
of algae, 75, 76
of benthic macroinvertebrates, 75–76
biological communities as indicators of ecological stress, 74
characterization of expected condition, 74
of fish populations, 74–75
index periods, 76
of lower trophic levels, 75–76
and reference sites, 74
Blanks, 81, 82*t.*
in identification of contamination, 87
Box plots, 159, 159*f.*

C

Cadmium, 19
Canada
Mine Environment Neutral Drainage (MEND) program, 2, 125, 177
sampling and monitoring guidance, 124*t.*, 125–126
Chain of custody, 73, 82
as cradle to grave record, 82
Chalk Creek (Colorado), 150
CIPW (Cross, Iddings, Pirsson, Washington) norm, 60
Clay, as confusing term (particle size or mineral group), 5–6
Climate change, 174
anticipating future of, 176
impact of mining operations on, 175
impact on mining operations, 175–176
types of, 174–175
Climate data, 16–17, 20
Closure, 25–26
defined, 25
and geomorphological stability, 25, 26*t.*
planning for, throughout mine life cycle, 1, 4
preliminary plans during mine development, 18
sampling and monitoring for objectives of, during operations, 24–25
sampling and monitoring needs, 26
sampling and monitoring programs, 4, 9, 10*f.*, 25–26
CMIST. *See* Multi Increment sampling tool
Colorado School of Mines, and mine waste decision tree (MWDT), 55
Combined Reserves International Reporting Standards Committee (CRIRSCO). *See* Committee for Mineral Reserves International Reporting Standards
Combustion-infrared analysis, 60
Committee for Mineral Reserves International Reporting Standards, 120
Comparability, 81
defined, 80, 80*f.*
Completeness, defined, 80, 80*f.*
Compliance monitoring, 24, 134
Computer aided drafting (CAD) software (in data management), 154
Concentrating, 22
Conceptual models, 136–137, 138*f.*, 139*f.*, 164
ASTM standard, 138
exposure pathways, 139
for numerical modeling, 140–142
purpose and use of, 138
receptors, 139, 139*f.*
reviewing existing data, 138
sources of acid drainage, 138
for water balance, 139–140, 140*f.*, 141*f.*
Confidence intervals, 162–163
Confidence level, 135–136, 136*t.*, 137*f.*
Confirmation bias, 160
Contamination, 86
blanks in identification of, 87
chemical preservatives as source of, 87
human-induced, 87
protecting against, 86–87
sources, 86–88, 87*t.*

Conventional coring, 72
Correlation, 162
Curvilinear regression, 162
Cyanide, 141

D

Data, 153, 169
  analysis and assessment of, 155–158, 169
  causal pathway assessments, 155
  components of management, 154
  condition assessments, 155
  and confirmation bias, 160
  contrasted with information, 160
  environmental assessments, 155
  evaluation of laboratory results, 155–158
  generation and acquisition, 78–79
  from global positioning systems (GPS), 154
  graphical representation of, 158–159
  integrating, 155
  interpretation, 160–164
  management, 153–155
  modeling, 164
  outcome assessments, 155
  predictive assessments, 155
  quality objectives, 79
  site assessments, 155
  spatial analysis, 163–164
  statistical analysis of, 161–163
  and uncertainty, 160–161
  and unreliability, 160
  unusual measurements and validity, 163
  usability, 83
  validation and usability, 79, 155–157, 157*f.*
  verification, 82–83, 83*t.*, 155–157, 157*f.*
  See also Climate data; Decision making; Environmental data; Quality assurance/quality control (QA/QC)
Data quality indicators (DQIs), 79–80, 80*f.*
Data quality objectives (DQOs), 35, 37, 38–39, 63, 67, 79
  as basis for planning sampling and monitoring programs, 133
  in determining objectives of sampling and monitoring programs, 135
  and disconnect with QA/QC and data verification, 83
Databases, 154
  defined, 154
  metadata, 154–155
  metadata required for standards and classifications proceedings, 156
  relational, defined, 154
  software for, 154
Decision Error Feasibility Trials (DEFT) modeling, 38–39
Decision making
  and adaptive management, 167
  and collaborative participation, 167–169
  identifying decision makers, 136
  and precautionary principle, 165–166, 167
  in public health and environmental risk, 166–167
  and sampling and monitoring data, 167
  sampling and monitoring programs as tools for, 153
  six-step process, 164–165
  See also Mine waste decision tree
Diel cycles, 44
  diel biogeochemical processes in streams, 45
Dissolved fraction, 40
Dose-response curve, 114
DQOs. *See* Data quality objectives
Drainage
  nonacidic, 4–5
  See also Acid mine drainage; Acid rock drainage; Acidic drainage; Mining influenced water
Drillhole information, 5
Drillhole/borehole sampling
  determining optimum drillhole spacing, 118–119
  drilling methods, 72
  methods, 71–72
  procedure, 70–71
  rock sampling, 71
  sampling plan, 70

E

Eh measurement, 41
Electron-probe microanalysis (EPMA), 58–59, 58*f.*, 59*f.*
Endangered species, 121
Environmental background, defined, 36
Environmental baseline, defined, 36
Environmental data
  during exploration, 13, 15–16

during mine development, 18, 20, 21
sampling and monitoring programs, 2–3, 5
See also Geo-environmental models
EPA. *See* U.S. Environmental Protection Agency
EPA and Hardrock Mining, 127
EPMA. *See* Electron-probe microanalysis
ESRI ArcGIS, 164, 169
European Union
and European Commission regulations of mines, 177
sampling and monitoring guidance, 123–125, 124*t.*, 125*f.*
EXAFS. *See* Extended X-ray absorption fine structure
Exploration, 9, 10*f.*
and climatic conditions, 16–17
comprehensive exploration and environmental sampling, 15–16
and disturbance of baseline conditions, 10–12
and environmental/economic considerations, 13, 15–16
and hydrologic features, 15
information needs, 9–10, 10*f.*
and landslide scarps, 14–15, 16*f.*
and MIW-related information, 9–10
nonrock environmental observations, 15
sampling and monitoring during, 13–17
as search for mineral deposit, 9
steps in (flowchart), 9, 11*f.*
Extended X-ray absorption fine structure (EXAFS), 57
Extraction methods, 61

F

Factor analysis, 162
Feasibility studies, 17
objectives, 14
Field leach test, 61
Filtration, 40–41
FoxPro (software), 154
Framework for Metals Risk Assessment, 113, 114, 127
Frequency distribution, defined, 162
Fundamental sampling error (FSE), 38, 64–66

G

Geo-environmental models (GEMs), 5, 18, 35–36, 89–90
Geographic information systems (GIS)
databases, 154
in spatial analysis, 164
Geomorphological stability, 25, 26*t.*
Geostatistics, 161, 164
Global Acid Rock Drainage Guide (GARD Guide), 3, 53, 127–129, 128*f.*, 177
Global positioning systems (GPS), 154
Global Reporting Initiative (GRI), G3 reporting guidelines, 111
Good neighbor agreement (Stillwater Mining Company), 168
Good Practice Guidance on Health Impact Assessment in the Mining and Metals Sector, 169
Groundwater
flow systems, 42, 43*f.*
Internet resources, 90
Groundwater samples, 50, 72
and aquifer heterogeneity, 50
and colloid formation, 50
and filtration, 50
and isotopes, 50–51
low-flow sampling, 50
and purging methods, 51–52
and representativeness, 51
and tracers, 50–51
and well design, 51
Grouping and segregation error (GSE), 38, 64–66
Guidance on Choosing a Sampling Design for Environmental Data Collection for Use in Developing a Quality Assurance Project Plan, 149
Gy, Pierre, 66
and Gy's sampling theory, 64

H

Handbook of Systems Engineering and Management, 119
Health Risk Assessment Guidance for Metals (HERAG), 114
Henderson mine (Colorado), and endangered species, 121

Histograms, 158, 159*f.*
Historical mines. *See* Legacy mines
Holden mine (Washington), and mineralogical studies, 20
Human health risk assessment (HRA), 114, 116
  cycle, 118, 119*f.*
Hydroacoustic meters, 47
Hyporheic zone, 42

I

Impact assessment monitoring, 134
Inactive mines. *See* Legacy mines
Increment delimitation error (IDE), 64–66
Increment extraction error (IEE), 64–66
Inductively coupled plasma–atomic emission spectroscopy (ICP-AES), 85
Inductively coupled plasma–mass spectroscopy (ICP-MS), 84
Integrated Risk Information System (IRIS), 114
International Council on Mining and Metals (ICMM), 114
  on partnerships in decision making for local communities, 169
  sustainable development framework, 111
International Finance Corporation, 177
Inventory monitoring, 134
IRIS. *See* Integrated Risk Information System
Iron Mountain mine (California), 163

K

Kensington Mine (Alaska), and discovery of acidic drainage in tailings dam construction, 12
Kriging, 164

L

Laser ablation inductively coupled plasma mass spectrometry (LA-ICP-MS), 60–61
Leaching, 22
Leadville, Colorado, 121
Legacy (historical, inactive, abandoned) mines, 23, 26–27
  and determining premining baseline conditions, 27
Linear regression, 162
Long-range heterogeneity error, 64

M

Matrix spikes, 81
Mean (arithmetic average), defined, 162
Mean deviation, 161–162
Median, defined, 162
MEND. *See* Mine Environment Neutral Drainage program
Metal Mining Technical Guidance for Environmental Effects Monitoring, 125–126
Metal toxicity, and bioavailability, 117, 117*f.*
Metallurgical testing, 5
Metals Environmental Risk Assessment Guidance (MERAG), 114
Methods for Chemical Analysis of Water and Wastes, 84
Methods of Soil Analysis, 56
MI sampling. *See* Multi increment sampling
Microbial sampling, 61–63
Milling, 22
Mine development
  and baseline environmental conditions, 18
  and climatological data, 20
  condemnation drilling, 18
  and environmental data, 20
  information needs, 18
  and metallurgical testing, 19
  mineralogical data, 20–21
  and operational plans, 18
  and permitting plans, 18
  and preliminary mine-closure plans, 18
  and trace elements, 19
Mine Environment Neutral Drainage (MEND) program (Canada), 2, 125, 177
Mine life cycle
  accounting for all aspects of, 1
  closure/postclosure, 9, 10*f.*, 25–26, 28
  exploration, 9–17, 10*f.*, 28
  mine development, 9, 10*f.*, 18–21, 28
  operations, 9, 10*f.*, 21–25, 28
  planning for closure throughout, 1, 4
Mine openings, 174
Mine planning software, 53
Mine Safety and Health Administration (MSHA), 146
Mine the Gap: Connecting Water Risks and Disclosure in the Mining Sector, 117

Mine waste decision tree (MWDT), 55
Mineral deposit, defined, 9
Mineralogy, mapping of, 61
Mining influenced water (MIW), 1, 2–3, 4
  common characteristics of, 39–40
  from lowering of water table, 22
  and mine openings, 174
  and nonacidic drainage, 4–5
  research needs and topics, 177–178
MLA, 60
ModAn, 60
Mode, defined, 162
Modeling, 164
  conceptual, 136–140, 138*f.*, 139*f.*, 140*f.*, 164
  conceptual models for numerical modeling, 140–142
  Decision Error Feasibility Trials (DEFT), 38–39
  empirical, 164
  geochemical, 49
  hydrogeochemical, 49
  mechanistic, 164
  predictive, 164
  total maximum daily load (TMDL), 142–143
  toxicological, 49–50
  uncertainty, 164
  and Visual Sample Plan, 38
Monitoring
  baseline, 5
  benefits of, 118, 119*f.*
  for closure objectives during operations, 24–25
  compliance, 24, 134
  defined, 4
  field, 29*f.*
  impact assessment, 134
  in operations, 24–25
  survey or inventory, 134
  trends, 134
  water balance, 25
  well, 28*f.*
  See also Sampling; Sampling and monitoring programs
MSHA. *See* Mine Safety and Health Administration
Multi increment (MI) sampling, 37, 66
  potential disadvantage of, 66–67
Multi Increment sampling tool (CMIST), 66
MWDT. *See* Mine waste decision tree

N

National Water Quality Monitoring Council, 39
New World mine (Montana), halted because of proximity to Yellowstone National Park, 15
New Zealand, sampling and monitoring guidance, 122*f.*, 123, 124*t.*
Normal distributions, defined, 162
Not safe zone, 114

O

Operations, 21
  excavation, 21–22
  milling, concentrating, and leaching, 22
  mine expansion, 23
  smelting and refining, 22–23
  standby or inactive mines, 23
Ophir Mine (Colorado), and avalanche effect on tailings, 172
Optical microscopy, 58–59, 58*f.*, 59*f.*
Ore deposit, defined, 9
Outliers, 162

P

Paradox (software), 154
Pathfinder elements, 15
PE. *See* Preparation error
Performing Quality Flow Measurements at Mine Sites, 43
Permitting
  and environmental data, 21
  and mine development, 18
  and mine expansion, 23
Periodic heterogeneity error, 64
PHREEQC aqueous geochemistry model, 141
Pierson correlation, 162
Portable X-ray fluorescence (PXRF), 57
Postclosure, 9, 10*f.*, 25–26
  defined, 25
  sampling and monitoring needs, 26
Potentially acid generating (PAG) rock, 70

Precision, defined, 79, 80*f.*
Prediction Manual for Drainage Chemistry from Sulphidic Geologic Materials, 53
Prefeasibility studies, 17
  objectives, 14
Premining conditions, 36–37
  defined, 36
Preparation error (PE), 64–66
Prospect sites, and acidic drainage, 12–13
Protocols
  in sampling and monitoring plans, 147–148
  for water sampling, 40
PXRF. *See* Portable X-ray fluorescence
Pyrite oxidation, and microbial activity, 63

Q

QEMSCAN, 60
Quality assurance project plans (QAPPs), 78, 126–127
  assessment and oversight, 79
  data generation and acquisition, 78–79
  data validation and usability, 79, 155–157, 157*f.*
  project management, 78
Quality assurance/quality control (QA/QC), 78, 90, 111, 155
  and accuracy, 81
  and analytical errors, 79
  and bias, 80–81
  and blanks, 81, 82*t.*
  and chain of custody, 82
  and comparability, 81
  and data quality indicators, 79–80, 80*f.*
  and data quality objectives, 79
  and data usability, 83
  data verification, 82–83, 83*t.*, 155–157, 157*f.*
  in defining confidence level for sampling, 135–136, 136*t.*, 137*f.*
  and disconnect with DQOs and data verification, 83
  and key error sources, 79
  and matrix spikes, 81
  and reference materials, 81
  and replicates, 81, 82*t.*
  samples, 40
  and splits, 81, 82*t.*
  and standard materials, 81
  and uncertainty, 80

R

R Project for statistical computing, 164, 169
Redox measurement, 41
Reference materials, 81
Refining, 22–23
Registration, Evaluation, and Authorization of Chemicals (REACH), 114
Regression analysis, 162
Relational databases, defined, 154
Remote sensing, 61
Replicates, 81, 82*t.*
Representative sample, 37–39
Representativeness, defined, 80, 80*f.*
Research needs and topics, 177–178
Resources and reserves reporting, 120, 121*f.*
Risk, 111, 112, 129
  Agricola on, 113
  classification of, in mining, 113, 113*f.*
  and consequences, 117, 118*t.*
  defined, 113
  and engineering systems, 120
  and environment, 115
  and human health, 114, 115
  and metals and metal compounds, 113
  of unavailable or incorrect data, 136, 137
  water-related, 117–118
Risk assessment, 114
  defined, 114
  environmental, 115–117
  and management, combined approach, 121–122
  and mine life cycle, 115–116
  process, for metals, 115, 116*f.*
  processes, compared, 114, 115*t.*
  terminology, 115*t.*
  variability vs. uncertainty in, 135, 136*t.*
Risk management, 115
  and assessment, combined approach, 121–122
  and data reliability, 120–121
  and mine life cycle, 115–117
  process, for metals, 115, 116*f.*
  terminology, 115*t.*

Risk Management Criteria for Metals at BLM Mining Sites, 127
Rock samples
  and chain of custody, 73, 82
  diamond drilling preferred for, 17
  drillhole/borehole sampling, 71
  storage of, 72–73
Roto-sonic coring, 72

S
Safe region, 114
Samples
  bias, 39
  contamination of, 86–88, 87*t.*
  defined, 4, 37
  deterioration of, 5
  preparation, 85–86
  preservation, handling, and storage, 88, 89*t.*
  quality assurance/quality control (QA/QC) samples, 40
  representative, defined, 37–39
  shift composite, 40
  and variability patterns (distributions), 135–136
  variation, 39
  See also Analytical chemistry; Groundwater samples; Rock samples; Solid samples; Surface-water samples
Sampling, 88–90
  analytical chemistry considerations, 84
  for closure objectives during operations, 24–25
  and compositional heterogeneity, 63–64
  comprehensive, of both exploration and environmental factors, 15–16
  data and media for, 34*t.*
  defined, 37
  defining questions, etc., 33–35, 88
  diamond drilling as preferred for rock samples, 17
  and distributional heterogeneity, 64
  and effects of heterogeneity, 63, 64*f.*
  environmentally related, timing of, 5
  error, 38
  error sources in, 63, 64
  fundamental sampling error (FSE), 38
  and GEMs, 35–36
  grids in systematic random sampling, 67, 68*f.*
  grouping and segregation error (GSE), 38
  and Gy's sampling theory, 64
  and heterogeneity of target population, 39
  information sources, 33
  Internet resources, 90–91
  issues with metals, 33
  and laboratory subsampling, 85–86
  and material strength properties, 19
  materials characterization, 25
  metallurgical, 19
  mineralogical, 20
  multi increment (MI), 37, 66–67
  in operations, 24–25
  planning, 33–35, 37
  and reduction and leaching operations, 22
  reliability of as determinant of analytical reliability, 63
  reproducibility and bias in, 67, 68*f.*
  and sample representativeness, 37–39
  and smelting and refining operations, 22–23
  sources of uncertainty, 69, 69*f.*
  and spatial and temporal scales of geologic systems, 65
  strategies and designs, 67–69
  stratified random sampling, 69, 69*f.*
  target population, 37, 39
  theory, 39
  Triad approach, 35
  uncertainty in, 39
  of water pumped from excavated areas, 21–22
  See also Air sampling; Biological sampling (aquatic); Drillhole/borehole sampling; Groundwater samples; Monitoring; Sampling and monitoring plans; Sampling and monitoring programs; Solid samples; Surface-water samples; Water sampling
Sampling and monitoring plans (SMPs), 145–146, 145*t.*, 151
  Chalk Creek example, 150
  communication strategy, 146–147
  defined, 146
  defining protocols, 147–148

defining standard operating procedures (SOPs), 147–148
field data collection information, 148
field equipment, 148
identifying decision makers, 136
implementation of, 151
and laboratories, 147
management, leadership, and organization, 146–147
and MSHA certification, 146
outline, 145*t.*
pilot studies, 150–151
and sampling design, 149–150
sampling parameters, 149–150
team qualifications and training, 147
Sampling and monitoring programs, 1
and additional data collection, 3–4
Australia and New Zealand, 122*f.*, 123, 124*t.*
Canada, 124*t.*, 125–126
and challenge of multiple phases over mine lifetime, 4
in closure/postclosure, 4, 9, 10*f.*, 25–26, 28
cost/benefit evaluations, 118–122, 119*f.*
data and sampling media for, 34*t.*
defining confidence level, 135–136, 136*t.*, 137*f.*
defining questions and objectives, 3, 133–136
developing site conceptual models, 136–140, 138*f.*, 139*f.*, 140*f.*
and DQOs, 133, 135
environmental, 2–3
and environmental goals, 134
European Union, 123–125, 124*t.*, 125*f.*
in exploration, 9–17, 10*f.*, 28
Global Acid Rock Drainage Guide (GARD Guide), 127–129, 128*f.*, 177
information sources, 3
and interaction with stakeholders, 167–169
international guidance documents and commonalities, 120–122, 122–129, 124*t.*
and legacy (historical, inactive, abandoned) mines, 26–27, 28
in mine development, 9, 10*f.*, 18–21, 28
in operations, 9, 10*f.*, 21–25, 28
in other countries, 177
planning process, 133–143
and QA/QC activities, 135
and risks of unavailable or incorrect data, 136, 137
steps in implementation of, 111, 112*f.*, 129
and sustainability, 111–112
as tools for decision makers, 153
types of, 134
United States (EPA), 126–127
and water balance, 134
See also Monitoring; Sampling
Scanning electron microscopy (SEM), 58–59, 58*f.*, 59*f.* *See also* Automated SEM-based quantitative mineralogical analysis
Scientific assessment, defined, 155
Securities and Exchange Commission (SEC). *See* U.S. Securities and Exchange Commission
SedNorm, 60
Selective-extraction techniques, 61
Semivariance, 164
Slopes
cut, 171
failure causes, 171–172
failures in mine-waste rock piles, 172
failures in tailings piles, 172
manufactured or filled, 171
Smelting, 22–23
SMPs. *See* Sampling and monitoring plans
Society for Mining, Metallurgy and Exploration (SME), 120
Soil Science Society of America, 56
Solid samples, 52, 89
and analytical methods, 60–61
and block model, 53
and bulk composition of, 57
classifying into geological categories, 52
and effects of heterogeneity, 63
geoavailability of, 57
heterogeneous nature of, 54
and homogenization and subsampling of, 54
method for waste rock and tailings, 52, 53
and microbial sampling, 61–63
and mine geologic model, 53, 89
mine waste decision tree (MWDT), 55

mineralogical characterization, 57–58
mining- and metallurgy-related, 52–54
multi-increment, 55
petrographic analysis, 58–59, 58*f.*, 59*f.*
and quantitative mineral determination, 60
sampling all lithologies, 54
sampling theory and errors, 63–67
sediment, 56
soil, 56
solid-extraction techniques, 61
statistically based sampling strategy, 54
See also Rock samples
Solid-extraction techniques, 61
SOPs. *See* Standard operating procedures
Spike, defined, 158
Splits, 81, 82*t.*
SPLP. *See* Synthetic precipitation leaching procedure
SQL Server, 154
Stability issues, 171
climate change, 174–176
mine openings, 174
monitoring needs, 171
research needs and topics, 177–178
slope stability and failure, 171–172
subsidence, 173–174
tailings impoundments, 172–173
Stakeholders
engagement with indigenous cultures, 177
and interaction with sampling and monitoring programs, 167–169
Standard deviation, 161–162
defined, 162
Standard materials, 81
Standard Methods for the Examination of Water and Wastewater, 84
Standard operating procedures (SOPs), 147–148
Standby mines, 23
Statistics, 161
confidence intervals, 162–163
geostatistics, 161, 164
terminology, 161–162
useful statistical quantities, 161
Stillwater Mining Company, good neighbor agreement, 168
STORET (water and biological database), 154, 169
Stratified random sampling, 69, 69*f.*
Subsidence, 173–174
defined, 173
sinkholes, 173
and water flow, 173
Sulfide minerals
as hosts to elements of concern, 59, 59*f.*
quantifying mineral phases influencing composition of drainage effluents, 59–60
Summitville mine (Colorado) climate data, 17
Surface-water samples, 44
discharge measurements, 47–49
modeling considerations, 49–50
and suspended sediment, 47
and synoptic sampling, 46, 48
temporal considerations, 44–47
total maximum daily load (TMDL) process, 48
tracer-dilution discharge, 48
and tracers, 49
Survey monitoring, 134
Sustainability, 111–112, 129
research needs and topics, 177–178
Synoptic sampling, 46, 48
Synthetic precipitation leaching procedure (SPLP), 61
Systematic random sampling, 67, 68*f.*

T

Tailings impoundments, 172–173
dam failures, 173, 174*t.*
release types, 172–173
slope failures, 172
"Tale of Two Mines," 136, 137
Target population, 37, 39
TCLP. *See* Toxicity characteristic leaching procedure
Terminology, 4, 127
differing definitions and interpretations of, 5–6
distinctions among drainage-related terms, 4–5
glossaries, 6

Test Methods for Evaluating Solid Waste, 84
Total maximum daily load (TMDL) models, 142–143
  water quality data needs, 142
Total maximum daily load (TMDL) process, 48
Toxicity characteristic leaching procedure (TCLP), 61
Toxicity field tests, 61
Tracer-dilution discharge, 48
Trend monitoring, 134
Trends (statistical), 162
Triad approach, 35
2007 SME Guide (mineral resources and reserves reporting), 120

U

U.S. Army Corps of Engineers Cold Regions Research and Engineering Laboratory, Multi Increment sampling tool (CMIST), 66
U.S. Environmental Protection Agency, 3
  methods for environmental sampling, 35
  monitoring-related publications, 127
  on soil sampling, 56
  quality assurance project plan (QAPP), 126–127
  Quality System and data quality objectives (DQOs), 35, 37, 38–39, 78, 126, 127*f.*
  sampling and monitoring program, 126–127
  sampling design guidance, 149
  STORET (water and biological database), 154, 169
U.S. Geological Survey (USGS), 3
  field leach test, 61
  and mine waste decision tree (MWDT), 55
  Mineral Resources On-Line Spatial Data Web site, 61
  risk management publications, 127
U.S. Securities and Exchange Commission (SEC), 120
Uncertainty
  aleatory, 163
  and data, 160–161
  epistemic, 163
  in modeling, 164
  and precautionary principle, 165–166
  and QA/QC, 80
  in sampling, 39, 69, 69*f.*
  vs. variability, 135, 136*t.*
United States, sampling and monitoring guidance, 126–127

V

Visual Sample Plan (VSP) software, 38

W

Water and Air Baseline Monitoring Guidance Document for Mine Proponents and Operators, 126
Water balance, 25, 42, 43, 134
  conceptual models of, 139–140, 140*f.*, 141*f.*
  and cyanide, 141
Water ionic balances, 157–158
Water monitoring
  in other countries, 177
  procedural assessment, 120, 122*f.*
Water sampling, 39, 72, 90
  and characteristics of MIW, 39–40
  clean procedures, 40
  data for TMDL models, 142
  field measurements, 41
  filtration in, 40–41
  and groundwater/surface-water interactions, 42, 43*f.*
  information sources, 39, 90
  passive devices, 42
  protocols, importance of, 40
  redox and Eh measurement, 41
  sensors, 41–42
  surface water, 44–50
  and water balance, 42, 43
Wire samplers, 40
Word processing software (in data management), 154
World Bank, 177
World Health Organization, 177

X

X-ray absorption near-edge structure (XANES), 57
X-ray absorption spectroscopy (XAS), 57
X-ray diffraction (XRD), 58, 60
  with whole-rock chemical analysis, 60